Long-range Low-power Devices Based on LoRa Backscattering for Next Generation IoT Applications

Long-range Low-power Devices Based on LoRa Backscattering for Next Generation IoT Applications

Marc Lazaro Marti
Antonio Lazaro Guillen
David Girbau Sala
Ramon Villarino Villarino

iSTE

WILEY

First published 2025 in Great Britain and the United States by ISTE Ltd and John Wiley & Sons, Inc.

ISTE Ltd
27-37 St George's Road
London SW19 4EU
UK

www.iste.co.uk

John Wiley & Sons, Inc.
111 River Street
Hoboken, NJ 07030
USA

www.wiley.com

Library of Congress Control Number: 2025941655

British Library Cataloguing-in-Publication Data
A CIP record for this book is available from the British Library
ISBN 978-1-83669-036-8

The manufacturer's authorized representative according to the EU General Product Safety Regulation is Wiley-VCH GmbH, Boschstr. 12, 69469 Weinheim, Germany, e-mail: Product_Safety@wiley.com.

Contents

Introduction

I.1. Why ultra-low-power long-range IOT matters

Envision a world where countless devices are connected to the Internet, constantly collecting, exchanging and processing data; not only conventional electronic devices such as computers, tablets or smartphones, but also everyday objects such as clothes, furniture, keys or even your wallet. This is the reality of the Internet of Things (IoT), a rapidly growing network of interconnected devices that is reshaping the way we live, work and interact with technology. From wearables that track our fitness to smart homes that automatically adjust the lighting and temperature, applications are unbounded. With billions of devices already connected and waiting for billions more in the coming years, the IoT is transforming industries ranging from healthcare to manufacturing, and promises to revolutionize our daily lives in ways we cannot yet imagine (Al-Sarawi et al. 2020; Yaici 2020).

At first glance, connecting mundane objects to the Internet may seem implausible to some, why add unnecessary complexity to everyday things? Collecting the temperature of clothes or detecting the humidity of a plant substrate can be interesting, but is it worth it? Does it really simplify our lives? It is certainly a controversial matter, and answers are always subjective. IoT nodes should always have a social–economic equilibrium, considering resources, costs, social benefits or security among others, and it is in this balance that the answer lies. A major critique of the IoT field arises from the use of devices that do not respect this balance, since they are usually designed from a flawed perspective, such as a marketing-driven approach, which unfortunately can lead to the creation of dysfunctional devices. However, it should be emphasized that the goal of the IoT is nothing other than improving the well-being of individuals, and never the opposite.

Sir Francis Bacon wrote in his work *Meditationes Sacrae – ipsa scientia potestas est* –, knowledge itself is power (Bacon 1859). Paraphrasing this in the 21st century, we could say that data is power, and in fact, IoT nodes are providing exceedingly large amounts of data, bringing insightful knowledge in many fields, which would be impossible to obtain through other methods.

The primary milestone for the next generation of IoT devices is to become imperceptible to users, requiring no battery replacements or periodic maintenance. IoT nodes must work seamlessly in the background, freeing users from having to be conscious of them, similar to how anti-lock braking systems (ABS) work in cars or how solar lights turn on and off automatically in smart cities. However, there are several challenges that need to be addressed for the IoT to reach its full potential (Bagchi et al. 2020; Firouzi et al. 2020; Souri et al. 2022):

– *Security:* one of the biggest challenges for the next generation of IoT will be to ensure the security of devices and data. As more and more devices are connected to the internet, there will be a greater risk of cyber-attacks and data breaches.

– *Interoperability:* a significant challenge is the lack of standardization between the wide variety of devices and technologies that are part of the IoT ecosystem, and ensuring that they can communicate and work together seamlessly.

– *Privacy:* the data collected by IoT devices can be highly sensitive and personal. With the growing amount of data being collected and analyzed by IoT devices, privacy concerns will become more critical than ever. Users will need to have control over their data and be confident that it is being handled securely and ethically.

– *Scalability:* the next generation of IoT will involve billions of devices and systems, which will require a scalable infrastructure and data processing capabilities.

– *Reliability:* IoT devices must be reliable and usually be capable of operating effectively in harsh conditions such as extended sun exposure or varying weather, without experiencing malfunctions or failures.

– *Complexity:* the IoT ecosystem is highly complex, with many different devices, protocols and technologies involved. Managing this complexity and ensuring that the IoT is easy to use and deploy is another challenge.

– *Energy efficiency:* many IoT devices are powered by batteries or other limited power sources, and optimizing their power consumption to extend their battery life is a major challenge. Devices will need to be designed to be energy-efficient and powered by sustainable sources.

– *Cost:* the cost of IoT devices and systems can be a barrier to adoption, especially in developing countries. To overcome this challenge, IoT technology needs to be made more affordable and accessible to everyone.

Great scientific efforts, both in public and private sectors, are underway to tackle each of these challenges. Although many of them are interconnected, each one of these challenges belongs to different fields and deserves its own independent research line.

For instance, while privacy must be addressed from the ethical–legal field, security must be addressed through engineering with robust communication protocols, and the optimal way to address interoperability is through standardization bodies. Addressing all of them in the same work would be an impractical and arduous task. Furthermore, it is crucial to acknowledge that a comprehensive solution must account for the interplay between hardware and software, as hardware alone cannot achieve its full potential without the support of suitable software measures.

There is currently a wide range of devices and communication protocols designed specifically for IoT (e.g. MQTT, CoAP, XMPP LoRa, ZigBee, NB-IoT, Z-wave, etc.). These protocols are intended to handle a large number of nodes ensuring compliance with the aforementioned requirements, while the devices are optimized to consume minimal amounts of energy. Unfortunately, in many cases, some of the requirements can only be met by sacrificing others, such as reducing energy consumption (El Alami and Najid 2020; Vinitha et al. 2022).

Despite continuous efforts to optimize transceiver power consumption, the power consumption of these devices remains orders above what is required to make them unnoticeable to users. Reducing energy demand below one milliampere is not an easy goal and is a crucial requirement to give a boost to the next generation of IoT devices. To reach this milestone, the hardware design must be approached from a completely different perspective. This is where the work presented in this book focuses.

I.2. A new approach: backscatter meets LoRa

Consumption, range and data rate are the three immovable pillars that govern wireless communications. Improving the performance of any one of them always involves, at least, sacrificing one of the other two. This trade-off must always be tailored to meet the requirements of each specific application.

The use of optimized radio frequency (RF) modulations, adjusting active operation cycles or biasing the circuit below the standard voltage are just some of the techniques used to deal with the aforementioned trade-off, improving one of the three properties while minimizing the negative effect on the others. Even so, the margin of action is limited, especially with commercial hardware.

In order to provide a clearer visual representation of the problem, Figure I.1 depicts the coverage of the main commercially available wireless technologies as a function of their power consumption. Two primary categories can be distinguished based on their power consumption: active and passive. Active transceivers cover the vast majority of technologies, starting from consumption levels of 10 mA and upwards. These consumption levels require the use of batteries in prototypes, increasing their manufacturing and maintenance costs. However, they are the only ones capable of providing long-range wireless communications and high data rates. Passive transceivers, on the other hand, operate in the range of μA, allowing devices to be powered using energy harvesting techniques or using alternative energy storage systems such as supercapacitors. The most common technology operating under this principle is radio frequency identification (RFID) (Ahson and Ilyas 2017).

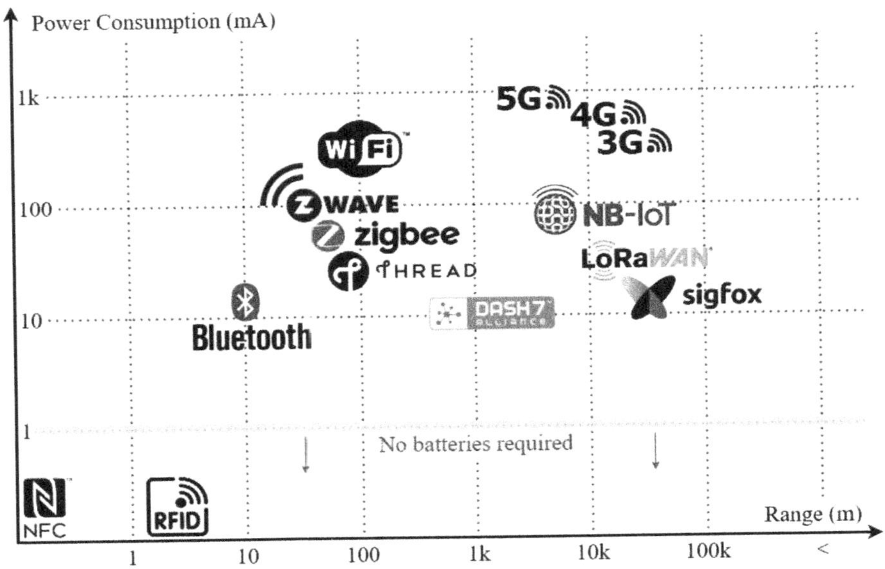

Figure I.1. *Current consumption as a function of read-range for different wireless technologies. For a color version of this figure, see www.iste.co.uk/lazaromarti/low-power.zip*

At first glance, we can observe that there is currently no commercial technology available to achieve wireless communications over tens or hundreds of meters while maintaining power consumption in the range of μA. This capability is an indispensable requirement for a subset of the next generation of IoT devices.

The work presented in this book focuses on the study of backscattering communications based on LoRa wireless technology for low-power, long-range links as an alternative to the recently exposed challenge. The backscattering communication is the basis of RFID and consists of reflecting RF waves instead of directly radiating them, reducing the consumption of the device by several orders.

RFID and backscattering communications have been limited to a few meters for several decades. However, with the advent of new, more complex modulations, such as spread spectrum, it has become possible to decouple backscattering communications from conventional RFID, which have historically relied on dedicated readers. This development has opened up a wide range of possibilities and led to very promising results. In the following chapter, backscattering communications will be thoroughly discussed.

The objective of this work is to explore the use of backscatter communications based on the LoRa wireless technology as a solution to the aforementioned gap within current available wireless technologies, providing a long-range low-power new wireless technology. This book aims to conduct a comprehensive review of the most significant contributions in the design of low-power, long-range backscattering systems. Subsequently, it seeks to make its own contribution in this field by designing a long-range low-power tag and validating its viability in real-world applications. Therefore, the main objectives are:

1) Introduce the basic principles behind backscatter communications and LoRa wireless technology, highlighting the advantages of LoRa over other modulations such as WiFi, Bluetooth or ZigBee.

2) Design and implement the tag front-end to maximize the backscattered power and, consequently, the range, but always maintaining the low power consumption constraint.

3) Propose the modulation scheme for the tag and a protocol for establishing the uplink communication, leveraging the key features of LoRa.

4) Evaluate the performance of the proposed tag in diverse applications within the medical, IoT and localization fields, thus validating its viability for future commercial applications.

I.3. How this book is organized

After a brief introduction to the IoT, where the main challenges in this field have been exposed, and the scope and objectives of this work have been focused, the structure of the book can be introduced, as well as the topics that will be addressed in each chapter:

– Chapter 1 begins with a historical introduction about the discovery and evolution of the backscattering technique, followed by a general review of its most important application fields: radar and RFID. Subsequently, a comprehensive review of the most relevant contributions to the design of bistatic backscatter communication systems is presented, covering both dedicated and ambient configurations.

– Chapter 2 focuses on conducting a theoretical review of the basic principles on the backscattering of electromagnetic fields. To achieve this, the fundamental theory of radar and antennas is reviewed. Furthermore, the basic principles of RFID are also reviewed to mathematically justify the effect of load switching on the antenna termination.

– Chapter 3 addresses the design and implementation of the backscattering tag. In this chapter, various topologies for the front-end design of the tag are discussed, along with two different modulations and communication schemes based on LoRa backscattering for transmitting the information from the tag. Furthermore, an introduction to LoRa modulation is made to expose the advantages and reasons for its use compared to other modulations. The following chapters present four different applications based on the LoRa backscattering technique described here.

– Chapter 4 studies the application of the backscattering technique in the field of localization. The proposed solution consists of a low-resolution localization system that leverages the low-power consumption of the backscattering tag and the high range provided by LoRa-based backscattering systems. System overview, proof of concept and comparison over other localization systems are provided.

– Chapter 5 proposes the design of an implanted device using the backscattering technique to establish wireless communication. This chapter focuses on studying the feasibility of using LoRa backscattering communications in high attenuation media, which is the case of the body.

– Chapter 6 presents the design of a smart mask that takes advantage of the low power of LoRa backscattering to implement the wireless communication. The mask integrates a dual-heat sensor capable of accurately measuring body temperature. Additionally, a simple localization system is proposed to determine whether the user wearing the smart mask has entered or exited a building.

– Chapter 7 presents the design of a home surveillance system in which wireless sensors are implemented using LoRa backscattering. This approach reduces the power consumption of the nodes, making them battery-free and consequently eliminating the need for periodic user maintenance. This simple use case demonstrates the potential of the technology covered in this book for smart home environments.

1

Background of Backscatter Communications

1.1. Discovery and evolution

Backscattering communications have a long and interesting history that dates back to the early days of radar in the 19th century. By then, James C. Maxwell had already published his work "A dynamical theory of the electromagnetic field" (Maxwell 1865), establishing the theoretical foundations of electromagnetic waves: Maxwell's equations. Maxwell's work was the culmination of a set of pieces of evidence presented by scientists such as Hans Christian Orsted, André-Marie Ampère and Michael Faraday during the 18th century that suggested a link between electricity and magnetism. Faraday showed that electricity could influence the behavior of a magnet and vice versa, planting the seed for Maxwell's work. The scientific community had been skeptical for more than a decade about the idea of Maxwell's fields until his theory was proven in 1888 by Heinrich R. Hertz (Cichon and Wiesbeck 1995). By the end of the 18th and early 19th centuries, wireless communications were becoming a reality thanks to the work of Hertz himself, Aleksander Popov, Nikola Tesla and Guglielmo Marconi with the creation of the first Morse code transceivers. In the first decade of the century, the work of Ernst F.W. Alexanderson and Reginals Fessenden made the first radio broadcast (Brodsky 2008) and amplitude modulation possible. At the dawn of World War I, wireless communications were still in the beginning stage and the main method of communication was wired. However, after the incessant destruction of communication lines, they soon realized the great advantage of wireless communications, and the four years of conflict were vital for the improvement of radio communications. The invention of the vacuum tube in 1904 by physicist John Ambrose Fleming (Dylla and Corneliussen 2005) and its subsequent industrial production in 1916 allowed radio transceivers to be miniaturized and integrated into planes, cars and boats. By 1918, with the end of the war, two-way radio communications and radio broadcasting were already a reality.

After the war, the study of electromagnetic fields was booming, and another major breakthrough was on the horizon: radio detection and ranging (radar). Years earlier, while attempting to prove Maxwell's theoretical work, Hertz discovered that radio waves were affected by metal objects and experimented with reflectors to observe this effect. The reflection of radio waves is known as backscattering, and it was the foundation for subsequently developing radar. Later, in 1904, Christian Hülsmeyer delved into Hertz's observations, determining that reflected radio waves could be used to detect distant metal objects, and he patented a naval detection system. The system, a predecessor to radar, did not attract much attention at the time. In 1917, Nikola Tesla established the theoretical principles of modern radar. In the 1930s, the pre-war environment in the face of the threat of World War II caused several countries to begin developing radar in parallel. The current radar model was created in 1935 in England by Robert Watson-Watt. However, Germany, the United States, Japan and the Soviet Union developed their own radars in parallel, primarily for military purposes, such as Freya, Wurzburg and SCR-268/270. With the outbreak of World War II, radar evolution progressed rapidly, driven by the war. The main problem with radar at that time was that it could not differentiate between enemy and allied targets. To solve this problem, the backscattered radar signals began to be modulated, giving rise to identification friend or foe (IFF) transponders. The race against time to win the war allowed for an increase in operating frequency, surpassing the pre-established barrier in the VHF band and giving rise to the first microwave radars. The main advantages of the frequency increase were the reduction in antenna size and the increase in precision due to narrow beam widths. One of the most important microwave radars was the SCR-584 (Getting 1990). After the war, although the development of radar slowed down, new advances such as the monopulse tracking radar, phased-array radar, synthetic aperture radar (SAR) and pulse Doppler radar continued (Skolnik 2008).

In 1945, World War II came to an end, marking the start of a new era: radio frequency identification (RFID). American inventor Leon Theremin created a completely passive listening device to spy on the Soviet Union. Although this device was not an identification tag but an audio modulation device, it is considered the predecessor of RFID because of its complete passivity, much like RFID tags. Three years later, in 1948, Harry Stockman published his work "Communication by means of reflected power" (Stockman 1948), which examined the theoretical basis of communications through backscatter. In his conclusions, Stockman stated the following:

> Evidently considerable research and development work has to be done before the remaining basic problems in reflected-power communication are solved, and before the field of useful applications is explored.

As expected, without any apparent military applications, it took more than two decades for several companies to show any interest in RFID. However, the development of RFID was already underway. In 1964, Roger F. Harrington presented a general formulation for electromagnetic backscatter in his work "Theory of loaded scatters" (Harrington 1964).

From 1970, the number of academic contributions in the field of RFID skyrocketed, and private entities began to show an interest in the potential applications of RFID tags in fields such as transportation, inventory control, animal tagging and personnel control. Mario W. Cardullo officially patented the first RFID tag with rewritable memory in 1973. Shortly thereafter, in 1975, scientists Alfred Koelle, Steven Depp and Robert Freyman presented a notable contribution titled "Short-range radio-telemetry for electronic identification using modulated backscatter" (Koelle et al. 1975). It described a simple and cost-effective electronic identification system capable of operating at distances of tens of meters, encouraging the large-scale commercialization of RFID. By the end of the 20th century, millions of RFID tags were being used around the world (Siegel 2021). It is worth mentioning that the evolution of RFID, like much of electronics, can be attributed to the invention of the transistor by William B. Shockley, John Bardeen and Walter Brattain in 1948 and its subsequent evolution into field-effect transistors (Landt 2005).

Currently, the field of backscattering continues to generate a lot of interest, not only for further improving existing technologies such as radar and RFID but also for developing modern backscattering communications. In the 21st century, scientific efforts have focused on addressing two of the main limitations of RFID: the imperative use of dedicated readers and the limited communication range, which had long been restricted to a few meters. In 2013, Joshua R. Smith et al. from the University of Washington presented their work "Ambient backscatter: wireless communication out of thin air" which proposed a communication system powered solely by RF signals from the environment. This breakthrough removed the first limitation of RFID, the need for dedicated readers. Over the last decade, researchers have successfully studied the use of various ambient signals as a carrier source and decoding backscatter communications using non-dedicated receivers. In 2017, the second limitation of RFID and backscatter communications was finally overcome with the introduction of the first long-range backscatter communication system based on LoRa signal reflection, with a maximum range of nearly half a kilometer (Peng et al. 2018). Today, the system has been further improved to enable communication at distances of over 1 kilometer (Jiang et al. 2021a). Figure 1.1 depicts the key discoveries and significant events in the evolution of backscatter communications.

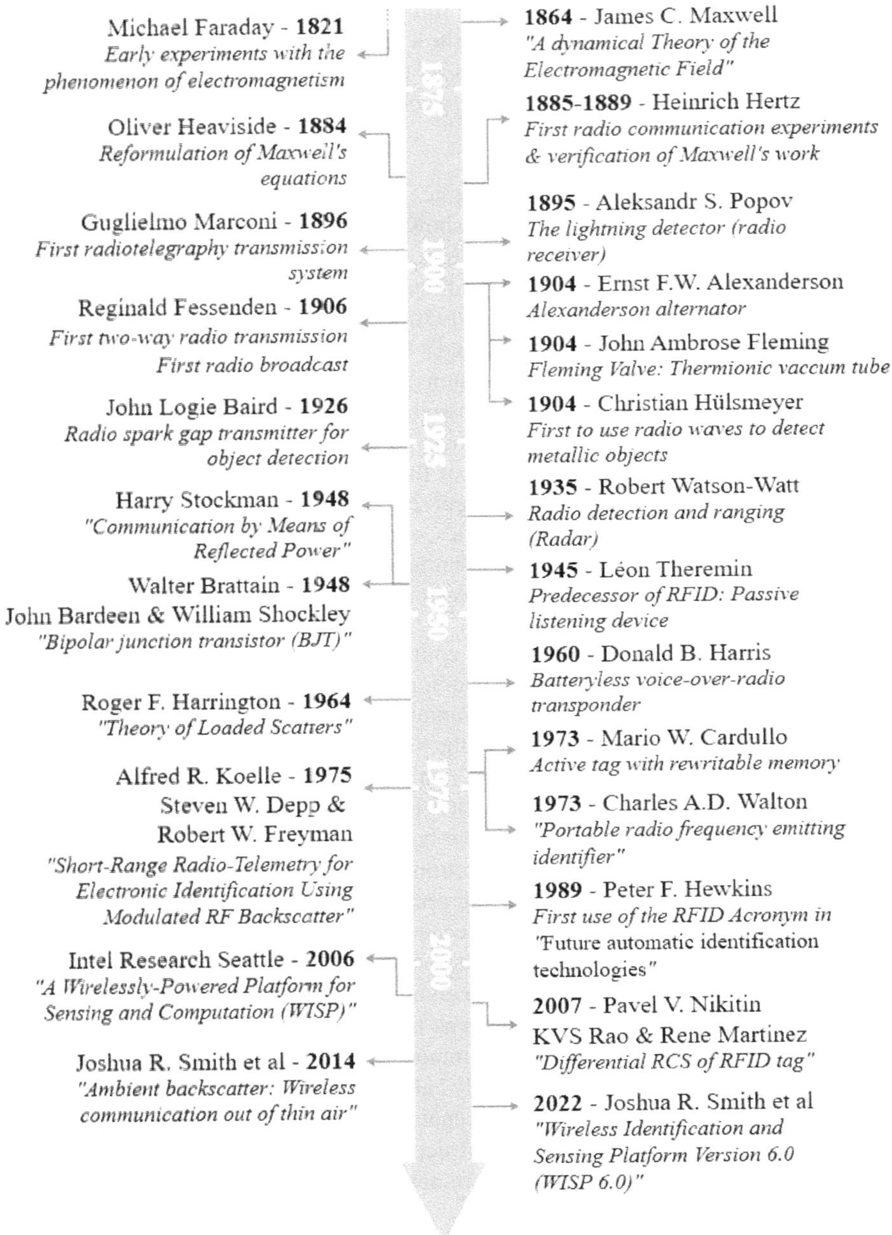

Figure 1.1. *Chronologic evolution of RF backscattering*

1.2. Modern backscatter communications

In this section, we review the current state of backscatter communications (BackCom), summarizing the latest research and technological advances. BackCom systems consist of three essential components: a carrier transmitter, a backscatter (modulator) and a backscatter receiver. These systems can be categorized based on various criteria such as topology, application, power requirements or modulation. One commonly used method for differentiation is based on the transmitter–receiver topology: monostatic or bistatic. Figure 1.2 shows the classification of the main BackCom systems and sets a guideline for the structuration of this section. Accordingly, the first and second sections will cover monostatic backscatter communication systems (MBCS) and bistatic backscatter communication systems (BBCS), respectively. BBCS will be analyzed more extensively than MBCS, since the scope of this book is more focused on this system architecture. Finally, the LoRa backscatter systems proposed in the literature will be evaluated in more detail in the last section.

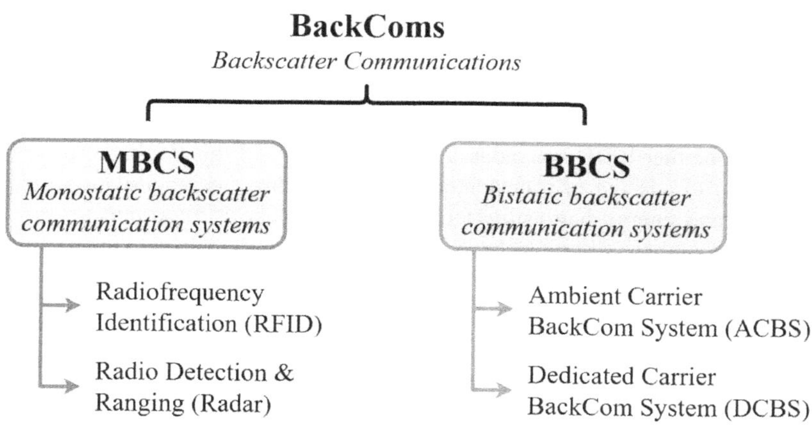

Figure 1.2. *BackCom system categorization. For a color version of this figure, see www.iste.co.uk/lazaromarti/low-power.zip*

1.2.1. *Monostatic backscatter communication systems*

Monostatic BackCom architectures integrate both carrier source and backscatter receiver in the same device, by definition called a transceiver[1], but usually known as a reader. Co-locating the carrier source and backscatter receiver introduces an undesired

1 Transceiver: device that can both transmit and receive signals for communication or detection purposes. In RFID, it is commonly referred to as a "reader". Radar is a transceiver.

self-interference to the system leading to what is known as a doubly near-far problem[2], limiting the communication range and adding extra complexity to the reader design. To cope with this problem, active cancellation techniques are required, increasing the cost of the reader (Sheikh et al. 2021). Moreover, in MBCS, signals suffer from a round-trip path loss (source-to-tag and tag-to-receiver) causing the signal-to-noise ratio (SNR) at the receiver to drop with the fourth power of the reader-to-tag distance (Kimionis et al. 2014d). When the tag is far from the reader, this particular loss translates into higher energy outage probabilities and a lower backscattered signal strength (Choi and Kim 2015), reducing the reader coverage.

The main backscattering technologies that use monostatic architectures are radar and RFID. Although both work on the same principle (obtain information from backscattered signals), there are significant differences between them. The main purpose of radar is to detect and track nearby objects without the need to add transponders to the target. In radar systems, the backscatter signal is not actively modulated, and target identification depends entirely on its radar cross-section (RCS). On the other hand, RFID involves adding a transponder[3], commonly referred to as a tag, to specific objects. These tags modulate the backscattered signal by altering one or several of its physical properties (amplitude, frequency or phase) in order to send data to the reader. Figure 1.3 shows an MBCS for both radar and RFID applications.

Although neither RFID nor radar are the main focus of this book, there are many similarities in the design of transponders, sensing techniques, modulation schemes, applications and energy harvesting techniques that make it important to provide a general review of these sibling technologies. Such a review can lead to insightful ideas that are of particular interest for the development of this and future research.

1.2.1.1. *Radar-based systems*

Research within the field of radar often focuses on enhancing transceiver capabilities, such as range and resolution, through the integration of cutting-edge components, the design of new topologies or the implementation of more advanced signal-processing algorithms. While these research fields are beyond the scope of this

2 The doubly near-far problem is a problem in wireless communication systems where a strong interferer signal causes the receiver to adjust its gain and sensitivity level, making it difficult to detect a weaker signal that arrives at the same time or shortly afterward. This problem occurs when the ratio between the power of the stronger and weaker signals is similar to the ratio between the power of the stronger and weaker noises in the channel. In such cases, the receiver may reduce its gain to avoid distortion from the strong signal, but this also reduces the ability to detect the weaker signal. This problem can occur in many types of wireless communication systems, including cellular networks, satellite communications and RFID systems.

3 Transponder: device that emits a signal in response to another receiving signal. In RFID, it is commonly referred to as a "tag".

book, there is a growing area of research within the radar field aimed at increasing the detectability of radar targets through the use of reflectors and active transponders.

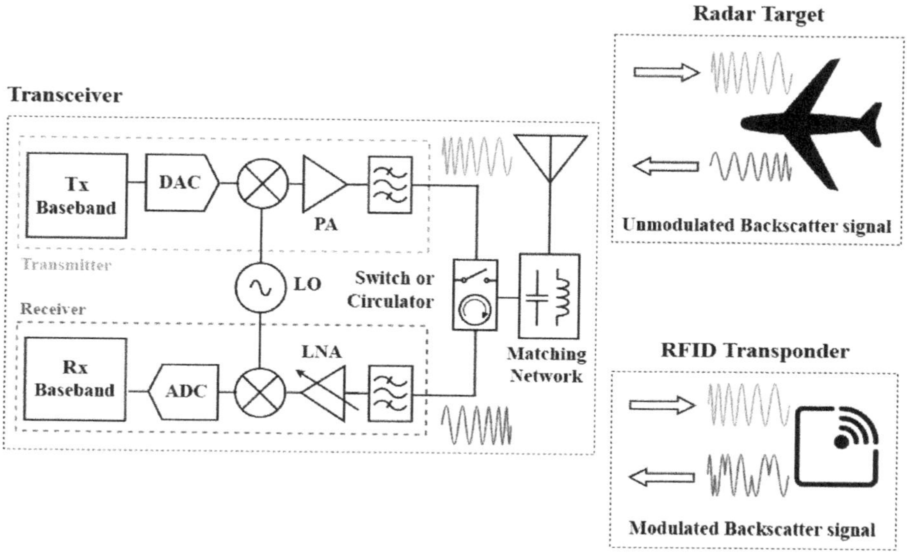

Figure 1.3. *Monostatic backscatter communication system topology. For a color version of this figure, see www.iste.co.uk/lazaromarti/low-power.zip*

Several contributions in the literature propose to extend the primary detection capabilities of radars by providing targets with backscattering-based transponders (Cnaan-On et al. 2014, 2015; Lazaro et al. 2014, 2015). The tag establishes an uplink toward radar to share data, such as an identifier or sensor value. Radars are used in a wide range of fields, and the most common ones can be classified into two main categories: continuous wave radar and pulsed radar. Among the most used ones we can find the frequency-modulated continuous wave (FMCW) radar, SAR and moving target indication (MTI) radar. The work in Fu et al. (2017) presents a mathematical model that can simultaneously image, localize and establish a one-way uplink communication between tagged objects and the radar using backscattering transponders. The proposed system uses SAR techniques and offers several advantages, including tag-versus-clutter discrimination, multiple access among tags and improved SNR in localization. Similar contributions are presented in Gu et al. (2017) and Venturino et al. (2022). With the relentless increase in the operating frequency of radars, transponder design for X-band (Lorenzo et al. 2015), Ku-band, K-band, Ka-band (Strobel et al. 2013) and mmWave-band (Stein et al. 2014) has generated great interest. High available bandwidth at the millimeter wave band is favorable for high-resolution ranging.

FMCW radars are widely used in the automotive industry to enable advanced driver assistance systems (ADAS). Radars excel at detecting metallic objects; however, their performance is poor when it comes to detecting pedestrians or objects made of other materials such as plastic or wood. For this reason, there is a growing research effort to increase the detectability of vulnerable targets that go unnoticed by automotive radars. Lazaro et al. (2020a) presents a microdoppler tag based on the signature of rotating backscatters, allowing the target to be detected in the presence of a strong clutter. In contrast to conventional active load-modulated transponders, this one is based on a rotating reflector. Recently, a double-antenna backscattering transponder has been proposed in the literature; this device can be used to generate spoofing attacks on FMCW vehicle radars (Lazaro et al. 2022a), establish inter-car communication (e.g. turn lights on, car stopped, car in reverse, etc.) (Lazaro et al. 2021a) and increase the detectability of vulnerable objects such as pedestrians, cyclists, road workers and, more recently, scooter riders, particularly in urban areas (Lazaro et al. 2023). The transponder includes an amplifier between both antennas to increase its RCS, and has been designed to operate in the 24 GHz band using both a pair of seven-element series-fed array antennas or a pair of patch antennas with two dielectric lenses.

Beyond the automotive sector, this tag-radar architecture is also considered in other fields of application. Ground penetrating radars (GPR) are used, among other purposes, to measure soil moisture. The main disadvantage is that the radar must be close to the ground or even in contact with it. Some studies have presented a hybrid system with backscattering tags to overcome this drawback (Josephson et al. 2019, 2021). Also in the medical field, various radar-based systems, such as CW Doppler and FMCW radars, are being studied to monitor respiration and vital signs with backscattering tags (Singh and Lubecke 2011; DiCarlofelice et al. 2020). Moreover, a similar configuration is used in Zhou et al. (2011) to implement an indoor localization system using an impulse radio ultra-wideband (IR-UWB) radar with low-power tags.

1.2.1.2. *RFID-based systems*

Parallel advancements in the design of both readers and tags have driven the progress of RFID technology. The main research lines have focused on enhancing techniques for processing the backscattered signal, enhancing reader sensitivity, reducing manufacturing costs, increasing the amount of data encoded/stored by the tag, and exploring new applications These research lines have been crucial in advancing RFID technology, making it more efficient and cost-effective, and expanding its potential uses in various fields. Beyond identification, RFID is currently used for diverse applications such as sensing, localization and authentication. Healthcare (Haddara and Staaby 2018), agriculture (Rayhana et al. 2021), transportation (Fong et al. 2019) and retail (Škiljo et al. 2021) are some of the main fields where RFID is used; however, the implementation fields of RFID are countless. For instance, in Nunes-Silva et al. (2019), RFID technology is evaluated as a system to study the behavior of bees. In recent years, RFID sensors based on polymers and nanomaterials have gained a lot of interest due to their capability to measure a wide variety of physical parameters such as temperature, humidity, pH, pressure, gases, biological compounds and mechanical stress (Costa et al. 2021). Figure 1.4 shows an example of an RFID sensor.

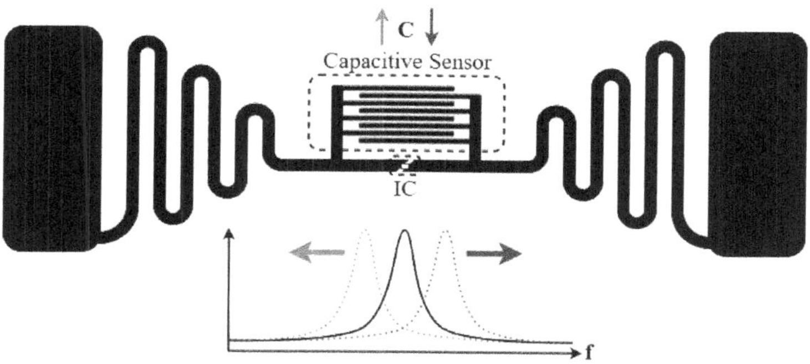

Figure 1.4. *Capacitive RFID sensor diagram. For a color version of this figure, see www.iste.co.uk/lazaromarti/low-power.zip*

One of the main advantages of RFID technology and the reason for its widespread adoption in countless applications is the low cost of tags. With the desire to definitively reduce the consumption of the tag to zero, chipless technology emerged, completely eliminating any type of electronics in the tag. Nowadays, RFID tags can be classified into two major groups: chipped and chipless. The former is the most commonly used in the commercial field, and its development is mainly driven by the private sector. Chipped tags can be classified according to several criteria, such as the frequency band where they operate (LF, HF, UHF, MW), power source (active, semipassive or passive), generation and performance of the integrated circuit (IC) (Higgs, Monza, UCODE, NTAG, etc.), or communication field (near-field or far-field).

On the other hand, chipless tags, technology that seeks to eliminate tag consumption entirely, are mainly classified by the domain where data is encoded: frequency domain, time domain, spatial domain or hybrid systems. In each of these subgroups, there exist several data codification techniques. The frequency domain is the first and most studied system in chipless literature, followed by the time domain. In the former, the most common technique consists of encoding information in frequency peaks through resonators (Mishra and Behera 2022), while in the latter, data is usually encoded by means of delay lines (Girbau et al. 2012; Forouzandeh and Karmakar 2015). Hybrid systems appear to cope with the encoding capacity limitations of the aforementioned techniques and consist of encoding information using two or even three domains, for example, acting on the frequency and time delay of the backscattered signal simultaneously (Vena et al. 2011; Babaeian and Karmakar 2018). The spatial domain, also known as the image domain, differs substantially from the others, as the tag's information is extracted by scanning the tag by sections (Nguyen et al. 2019). This technique heavily depends on the resolution of the reader and is therefore usually only used at frequencies above 10 GHz. Figure 1.5 shows the classification of RFID transponders.

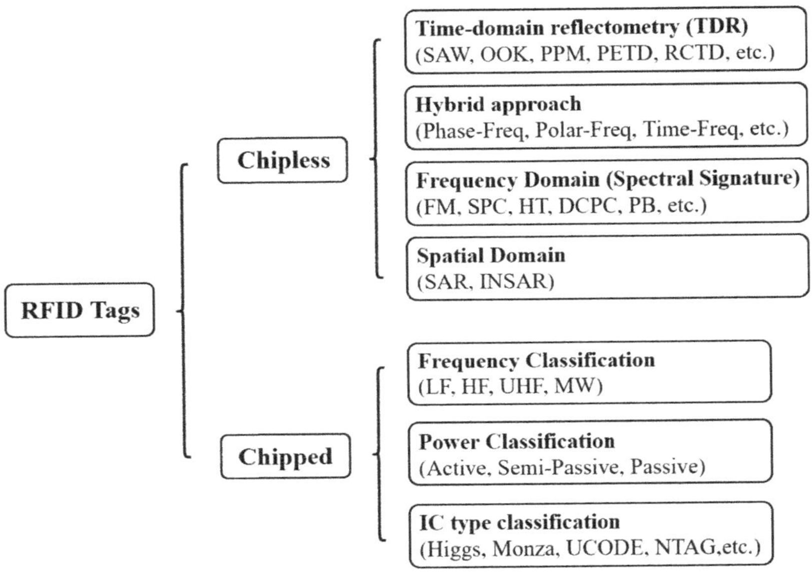

Figure 1.5. *RFID transponder classification*

In 2006, a significant breakthrough in the transponder side design was achieved with the introduction of the wireless identification and sensing platform (WISP). This innovation resulted from a collaboration between Intel Research Seattle and the University of Washington (Smith et al. 2006). Prior to the development of WISP, RFID tags were limited to being non-programmable black boxes, preventing researchers from conducting experiments with them. WISP revolutionized the field by introducing the first battery-free embedded transponder that integrated an MSP430 microcontroller, enabling it to be read and powered by commercial UHF readers. Over time, the WISP has undergone substantial enhancements, improving its performance and incorporating various low-power sensors capable of operating solely on harvested energy from RF signals and ambient light. Figure 1.6 shows the latest version of the WISP device released in 2022 (Menon et al. 2022).

Another important branch of RFID is near-field communication (NFC) technology. The typical operating frequency of NFC is 13.56 MHz, within the high-frequency

(HF) band. Unlike conventional RFID, NFC uses inductive coupling as a means of communication and energy transmission, limiting the communication range to a few centimeters. NFC devices use coil antennas to concentrate the magnetic field and maximize the coupling between the two devices. The most well-known application of NFC is contactless cards, initially for bank cards, but nowadays also used in identification cards such as national identity documents or public transportation cards.

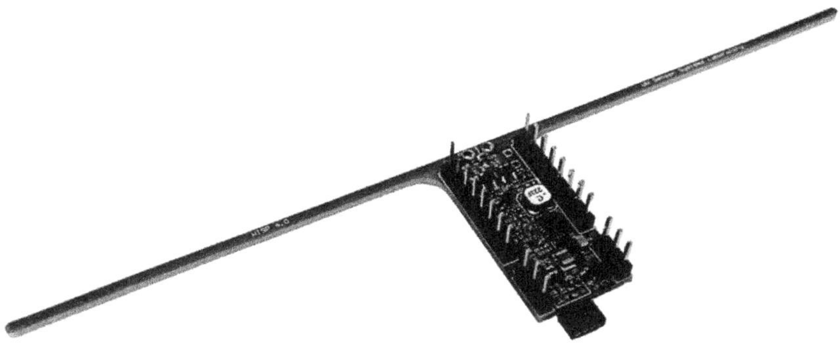

Figure 1.6. *Wireless identification and sensing platform (WISP) version 6.0 (Menon et al. 2022). For a color version of this figure, see www.iste.co.uk/lazaromarti/low-power.zip*

The recent integration of this technology into smart devices has encouraged the development of passive NFC tags in a wide range of fields, opening the door to NFC sensors (Boada et al. 2022). Some of the most recent contributions in the literature propose novel applications such as a battery-less bicycle tire pressure sensor based on a force-sensing resistor (Boada et al. 2021), a flexible epidermal NFC tag for sensing sweat (Barba et al. 2022) and a battery-less NFC potentiostat for measuring glucose strips and other electrochemical point-of-care sensors (Lazaro et al. 2022b).

1.2.2. *Bistatic backscatter communication systems*

In contrast to MBCS, in bistatic architectures, the carrier source and backscatter receiver are deployed separately as described in Figure 1.7. BBCS adds certain advantages to the system compared with MBCS, the reason why it has recently been adopted as the architecture for next-generation long-range backscattering systems. The individual design of the carrier source and the backscatter receiver reduces their complexity, and consequently their cost. In most scenarios, losses in BBCS systems never reach the round-trip path loss. Only when the tag is equidistant from the transmitter and the receiver, will the system exhibit the same losses as in the monostatic configuration. However, considering that the tag does not have a static location, in most cases, it will not fulfill the aforementioned condition. Moreover, the doubly near-far problem can be mitigated by dislocating the carrier source and backscatter receiver, and even more, deploying multiple carrier sources, allowing remote targets to harvest enough energy from nearby power beacons to reach the minimum required power level at the backscatter receiver. The strategic placement of carrier emitters can greatly enhance the coverage of the system (Jia and Zhou 2021).

However, compared with its monostatic counterpart, the dislocating carrier source and backscatter receiver introduce a new problem to be addressed: a carrier frequency offset (CFO) between the source and the receiver. This miss-synchronization severely affects the system throughput (Kimionis et al. 2014d) and is especially problematic for low-data rate communications. Historically, periodogram or preamble techniques have been proposed to compensate for the CFO problem (Fasarakis-Hilliard et al. 2015). However, more recent research proposes a novel interface cancellation scheme for BBCS that does not suffer from CFO, avoiding the long processing times that periodogram techniques can require in some scenarios (Tao et al. 2021).

Within BBCS, there are two major groups: those that require a dedicated deployment and those using ambient signals from legacy devices, also known as ambient backscatter communication (AmBC) systems. Subsequently, an overview of the principal contributions in both fields will be examined.

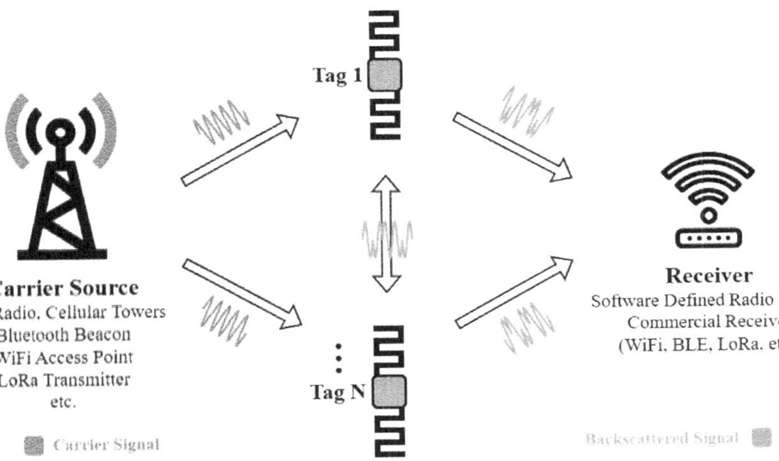

Figure 1.7. *RFID transponder classification. For a color version of this figure, see www.iste.co.uk/lazaromarti/low-power.zip*

1.2.2.1. *Dedicated carrier-based systems*

Despite the appeal of using ambient signals as a carrier and power source for backscatter communications, this configuration is not always feasible, especially in areas with little or no coverage, such as high mountain zones or garages. Dedicated carrier-based systems can not only cope with these situations, but may even be a preferable option over AmBC systems to meet certain coverage, reliability and performance requirements. In contrast to AmBC, where transmitters are fixed in specific locations, the use of a dedicated carrier source adds control over an extra variable in the system design, the transmitter, which results in several advantages. Carrier source location can be carefully determined as a function of the environment to maximize system performance, and multiple carrier sources can be deployed, providing coverage to dead zones, which would be impossible with a single carrier source. In addition, the use of dedicated carrier sources gives control over parameters such as the output power and carrier modulation, avoiding the problem of modulating the tag information onto already modulated and time-varying signals, making it easier for the receiver to decode the information.

BBCS based on multiple dedicated carrier sources (also called power beacons) has been extensively studied in the literature to increase system coverage (Kimionis et al. 2012). The placement of carrier sources is not trivial, and an appropriate arrangement can maximize the guaranteed coverage distance (GCD) (Jia and Zhou 2021). Additionally, by using multiple carrier emitters, transmitted power can be reduced by a factor of 50 (Kimionis et al. 2013). A recent contribution proposes the addition of an intelligent reflecting surface (IRS) to assist tag-to-reader backscatter communication: phase changes introduced by the IRS are addressed by implementing a miniaturization–maximization algorithm, and a multiple-tag scenario is considered (Jia et al. 2021).

While there is plenty of research work successfully addressing the main challenges of BBCS through novel coding schemes, multi-antenna techniques, anti-collision protocols, channel modeling, etc., the implementation of these systems with dedicated devices typically needs to be done from scratch on general-purpose equipment such as FPGAs. This approach can be expensive and discourages large-scale commercial adoption. As a result, some contributions propose hybrid systems compatible with existing wireless modulations and commodity receivers but still using dedicated carrier sources, starting the transition to ambient backscattering architectures. Some of the most relevant contributions of BBCS based on a dedicated carrier emitter are summarized in Table 1.1.

A WiFi BBCS is presented in Kellogg et al. (2016). In the proposed system, a dedicated carrier source emits a frequency tone that the backscatter transponder uses to directly synthesize 802.11b packets, later demodulated by commodity WiFi receivers. The idea was tested experimentally with an FPGA, and power consumption was simulated with the design of an application-specific integrated circuit (ASIC). The results stated a power consumption of 14.5 μW and 59.2 μW for 1 $Mbps$ and 11 $Mbps$ data rates, respectively. A comparable implementation, which is compatible with standard Bluetooth receivers, is outlined in Ensworth and Reynolds (2015). The proposed backscatter tag effectively synthesizes BLE advertisement packets (unidirectional uplink communication) from a CW carrier signal. The backscattered BLE packets are indistinguishable from those generated by a BLE transmitter, maintaining the same channel, modulation scheme and packet structure, and neither hardware nor firmware alterations are required to enable commodity BLE devices to demodulate the backscattered signal. The experimental results demonstrate communication ranges of up to 9.4 m with a data rate of 1 $Mbps$, providing performance similar to that of active BLE radios but with significantly lower power consumption.

Ref	Carrier	Receiver	Performance		
			Consumption	Throughput	Tag-Rx
Kimionis et al. (2012) Kimionis et al. (2013) Kimionis et al. (2014d)	Frequency Tone (867 MHz)	Custom (SDR:USRP)	n/a	1 $kbps$	134 m (Tx: 13 dBm)
Ensworth and Reynolds (2015)	Frequency Tone (2.4 GHz)	Commodity (BLE)	28 nW (FE) Ctrl(FPGA)	1 $Mbps$	9.4 m
Fasarakis-Hilliard et al. (2015)	Frequency Tone (868 MHz)	Custom (SDR:USRP-2)	n/a	1 $Mbps$	150 m 13 dBm
Kellogg et al. (2016)	Frequency Tone (2.4 GHz)	Commodity (WiFi)	14.5 - 59.2 μW (IC)	1 $Mbps$ - 11 $Mbps$	9.1 - 30.5 m
Pérez-Penichet et al. (2016)	Frequency Tone (2.4 GHz)	Commodity (ZigBee)	n/a (FPGA)	250 $kbps$	20 cm 0 dBm
Talla et al. (2017a)	Frequency Tone (905 MHz)	Commodity (LoRa: SX1276)	9.25 μW (IC)	18 bps - 37.5 $kbps$	237.5 m (2.8 km, Tx-Tag: 5 m)
Varshney et al. (2017)	Frequency Tone (CC2420) (0.86-2.4 GHz)	Custom (CC1310-868) (CC2500-2.4)	70 μW (868) 650 μW (2.4)	2.9/197 $kbps$ (868/2.4)	3.4 km (868) 225 m (2.4)
Hessar et al. (2019)	Frequency Tone (USRP+UBX-30) (900 MHz)	Custom (USRP+UBX-40)	45.2 μW (IC)	1 $kbps$	n/a
Katanbaf et al. (2021)	Frequency Tone (ADF4351) (900 MHz)	Custom (Tx-Rx co-loc) (SX1276+ADF)	(Receiver) 112 mW(30dBm) 3.04 W (4dBm)	366 bps - 13.6 $kbps$	103 m

Table 1.1. *Comparison between the most relevant contribution in BBCS based on dedicated devices*

Likewise, the system design mentioned above can be extrapolated to many other modulations such as ZigBee (802.15.4) (Pérez-Penichet et al. 2016) or LoRa (Talla et al. 2017a; Varshney et al. 2017). LoRa, which stands for long-range, is a widely adopted wireless technology in the Internet of Things (IoT) that uses chirp spread spectrum (CSS) modulation to encode information. CSS modulation offers several advantages, including high resistance to interference and high processing gain, resulting in extended communication ranges. As a result, backscatter communications based on this modulation leverage these properties to enhance the coverage of the backscattering system. Research studies such as LoRea (Varshney et al. 2017), LoRa backscatter (Talla et al. 2017a) and full-duplex LoRa backscatter (Katanbaf et al.

2021) report communication ranges of up to 3.4 km, 2.8 km and 92 m, respectively, under specific conditions such as the tag-to-transmitter distance, the gain of the antennas, the power of the transmitter and the type of scenario (indoor/outdoor).

As research tends to use ubiquitous commodity devices in dedicated or ambient BBCS, the line that distinguishes both groups becomes blurred. Finding a classification criterion that applies to any backscattering system seems difficult, if not impossible. Some works consider any architecture with no control over the transmitter as ambient BBCS, such as those based on TV or FM carriers. On the other hand, others consider any system that does not require the deployment of dedicated transceivers as ambient BBCS, such as those based on ubiquitous commodity transceivers. Note that the main difference between these two groups is the control over the carrier signal. In this work, any architecture in accordance with the second definition has been considered an ambient BBCS and assessed in the following section. Table 1.1 summarizes the stated characteristics of the most relevant contributions to dedicated BBCS found in the literature.

1.2.2.2. *Ambient carrier-based systems*

Joshua R. Smith and Alanson P. Sample, the two main researchers in the WISP project, conceived the idea of powering tags with energy harvested from ambient RF signals, thereby freeing the system from the imperative use of conventional RFID readers. Despite the bulky antenna used in preliminary tests, they successfully collected $60\mu W$ at a range of 4 km from a TV tower. As a proof of concept, a commercial temperature and humidity sensor was powered up (Sample and Smith 2009). Later, the harvesting sensitivity of the system was enhanced, allowing sensor node operation at a distance of 10.4 km from a TV transmitter and over 200 m from a cellular base transceiver station (Parks et al. 2013). An important milestone was reached in 2013 with the first ambient backscatter communication (AmBC) system presented by Liu et al. (2013a). This innovative system enabled inter-communication between several backscattering tags using TV transmissions as the sole source of power, without the need for specifically deployed transmitters or receivers. The proposed tags integrated a backscatter transmitter and receiver, allowing communication among them at distances up to 0.76 m with a predefined data rate of 0.1, 1 or 10 $kbps$ (Liu et al. 2013a). Since then, research on AmBC systems has gained momentum. In 2014, an improved version of the backscatter transmitter and receiver, named μmo and $\mu code$, respectively, was introduced in (Parks et al. 2014). The proposed receiver (μmo) implemented a low-power multi-antenna design to separate the noisy TV transmission from the backscatter transmitter data, resulting in a 100-fold increase in data rate, from 10 $kbps$ to 1 $Mbps$, and an extended range up to 24 m. On the other hand, for the design of the $\mu code$ transmitter, a novel coding scheme that eliminates the need for transmitter–receiver synchronization was proposed, thereby reducing the complexity of demodulating tag information at the receiver side.

Ref	Carrier	Receiver	Performance		
			Consumption	Throughput	Tag-Rx
Ishizaki et al. (2011)	WiFi (2.4 GHz)	Commodity (WiFi)	30.8 μW	5.5 $kbps$	0.3 m
Liu et al. (2013a)	TV (539 MHz)	Tag[4]	0.79 μW (FE)	10 $kbps$	2.5 m
Parks et al. (2014)	TV & RFID (539 & 915 MHz)	Tag	(μcode) 8.9 μW (μmo) 422 μW	(μcode) 1 $kbps$ (μmo) 1 $Mbps$	(μcode) 24.3 m (μmo) 2.13 m
Kellogg et al. (2014)	WiFi (2.4 GHz)	Commodity (WiFi)	9.65 μW (FE)	1 $kbps$	2.1 m
Bharadia et al. (2015)	WiFi (2.4 GHz)	Commodity (WiFi)	3.15-15.75 μW	5 $Mbps$ (1 $Mbps$)	1 m (5 m)
Zhang et al. (2016b)	WiFi (2.4 GHz)	Commodity (WiFi)	n/a	100 $kbps$ (1 $Mbps$)	50 m (8 m)
Zhang et al. (2016c)	WiFi (2.4 GHz)	Commodity (WiFi)	33 μW (IC)	300 $kbps$ (LOS) 200 $kbps$ (NLOS)	54 m (LOS) 32 m (NLOS)
Iyer et al. (2016)	BLE (2.4 GHz)	Commodity (WiFi & ZigBee)	28 μW (IC)	2 $Mbps$ (up to 11 $Mbps$)	27.4 m
Zhang et al. (2016a)	WiFi & BLE (2.4GHz)	Commodity (WiFi & BLE)	45 μW (IC)	313.8 bps (WiFi) 45.8 bps (BLE) 48.7 $kbps$ (FS)	4.8 m (WiFi) 4.4 m (BLE) 3.6 m (FS)
Wang et al. (2017)	FM (88-108 MHz)	Commodity (FM)	11.07 μW (IC)	3.2 $kbps$	18.2 m
Daskalakis et al. (2017a)	FM (88-108 MHz)	Commodity (FM)	313 μW	500 bps	2 m
Daskalakis et al. (2017b)	FM (88-108 MHz)	Commodity (FM)	0.677 mW (2.838 mW)	50 bps (2.5 $kbps$)	5 m
Vougioukas and Bletsas (2017)	FM (88-108 MHz)	Commodity (FM)	24 $\mu W(1.2\ V)$ 58 $\mu W(1.79\ V)$	Analog frequency modulation	15 m (indoor) 26 m (outdoor)
Zhang et al. (2017)	WiFi, BLE & ZigBee (2.4 GHz)	Commodity (WiFi, BLE, & ZigBee)	30 μW	60 $kbps$ 15 $kbps$ (multi)	42 m
Yang et al. (2017)	FM, TV & Cellular (88-900 MHz)	Custom (SDR)	n/a	158 bps	22 m
Talla et al. (2017b)	Basestation	Tag (AM)	3.48 μW	n/a (AM)	9.4 m (RF) 15.2 m (Light)
Vougioukas and Bletsas (2018)	FM (88-108 MHz)	Commodity (FM)	24 μW	FM mod P-FSK (1 $kbps$) S-BPSK (0.5 $kbps$)	15 m (indoor) 26 m (outdoor)
Zhao et al. (2018a)	WiFi (2.4 GHz)	Commodity (WiFi)	n/a	50 $kbps$	14 m
Zhao et al. (2018b)	WiFi (2.4 GHz)	Commodity (WiFi)	14.2 mW (FPGA)	200 bps	8 m (tag-tag: 0.4 m)
...					

Table 1.2. *Comparison of AmBC systems proposed in the literature (Part 1)*

4 Tag integrates both transmitter and receiver capabilities, allowing tag intercommunication.

Ref	Carrier	Receiver	Consumption	Throughput	Tag-Rx
		...			
Abedi et al. (2018) Abedi et al. (2020)	WiFi (2.4 GHz)	Commodity (WiFi)	10 μW	4-40 $kbps$	8 m (20 m)
Peng et al. (2018)	LoRa Gateway (900 MHz)	Commodity (LoRa)	220 μW	6.25 $kbps$ (SF7) 97 bps (SF12)	300 m (SF7) 800 m (SF12)
He et al. (2019) He et al. (2020)	WiFi (2.4 GHz)	Custom (SDR: USRP)	n/a (FPGA)	700 $kbps$	0.6 m
Kapetanovic et al. (2019)	SDR (WiFi, FM, TV) (91-2400 MHz)	Commodity (WiFi, FM, TV)	800 μW 50 μW (listen)	10 $kbps$	10 m (FM) 9 m (TV) 3.5 m (WiFi)
Guo et al. (2020)	LoRa (900 MHz)	Custom (SDR: USRP)	300 μW	9.9 - 121.4 $kbps$	71.4 - 225.6 m
Zhang et al. (2020a)	BLE (2.4 GHz)	Commodity (BLE)	n/a (FPGA)	2.8 $kbps$	23 m
Zhang et al. (2020b)	BLE (2.4 GHz)	Commodity (BLE)	37 μW (IC)	16.6 $kbps$	25 m (indoor) 56 m (outdoor)
Liu et al. (2020a)	WiFi (2.4 GHz)	Commodity (WiFi)	464 μW 32 μW (IC)	240 $kbps$ (single) 246 $kbps$ (multi)	21.3 m
Gong et al. (2020)	WiFi, BLE & ZigBee (2.4 GHz)	Commodity (WiFi, BLE & ZigBee)	279 mW	278.4 $kbps$	28 m (WiFi) 22 m (BLE) 20 m (ZigBee)
Jung et al. (2020)	ZigBee (2.4 GHz)	Commodity (WiFi)	2 mW (FPGA) 38.7 μW (IC)	200 $kbps$	27 m
Chi et al. (2020)	LTE (680 MHz)	Custom LTE Rx (SDR: USRP)	>4.5 mW 158.6 μW	3.63 $Mbps$	97.5 m
Zhang et al. (2021)	BLE (2.4 GHz)	Commodity (BLE)	2.03 mW (FPGA)	8.275 $kbps$ (LOS)	20 m (LOS) 14 m (NLOS)
Wang et al. (2021)	WiFi (2.4 GHz)	Custom & Commercial (SDR: USRP)	n/a	237.8 $kbps$	16 m
Dunna et al. (2021)	WiFi (2.4 GHz)	Commercial (WiFi)	30 μW (IC)	500 $kbps$	30 m
Liu et al. (2021)	WiFi (2.4 GHz)	Custom WiFi (SDR: USRP)	30.2 μW (IC)	62.3 $kbps$ - 10.61 $Mbps$	48.76 m
Xu and Gong (2021)	ZigBee (2.4 GHz)	Commodity (ZigBee)	n/a (FPGA)	13 $kbps$ (LOS) 10.3 $kbps$ (NLOS)	15 m
Jiang et al. (2021a)	LoRa (433 MHz)	Custom (SDR:USRP N210)	320 μW	120 bps	2.2 km
Guo et al. (2021)	LoRa (900 mHz)	Custom (SDR:USRP N210)	Receiver: 0.3 mW	39.5–199.4 $kbps$	50-200 m
Guo et al. (2022)	LoRa (433 MHz)	Custom (COTS design)	Receiver: 93.2 (IC)	19.6 $kbps$	180 m
Xu and Gong (2023)	BLE (2.4 GHz)	Commodity (ZigBee)	n/a	218 $kbps$ (LOS) 204 $kbps$ (NLOS)	24 m 20 m (BER<1%)
Liao et al. (2023)	LTE (R&S SMBV100) (486 MHz)	Custom (SDR:USRP)	n/a (Rasp. Pi)	n/a	21 m (LOS)

Table 1.3. *Comparison of AmBC systems proposed in the literature (Part 2)*

5 Analog frequency modulation refers to a modulation technique where there is no digital data. The signal varies based on a physical magnitude, such as a capacitive sensor. Analog modulation cannot be directly compared to digital modulations in terms of bits per second (bps). However, an approximate bps value can be extrapolated based on the resolution of the analog demodulator receiver.

WiFi transmissions are also proposed as a potential technology to harness backscatter communications. Nowadays, it is difficult to find an urban area without WiFi networks. Smartphones, wearables, tablets and laptops all integrate WiFi capabilities, making it highly appealing to be used in AmBC systems, not only as carrier emitters but also as receivers. The WiFi AmBC system was first proposed in Kellogg et al. (2014). The novel approach proposed a full duplex communication capable of connecting RF-powered tags to the internet by means of commodity WiFi devices. The uplink communication is implemented by modulating the existent WiFi channel information, while the downlink communication is implemented with the presence or absence of WiFi packets. Results stated a communication range of up to 2.1 m and a data rate of up to 1 $kbps$. A similar approach is presented by Ishizaki et al. (2011). The proposed system directly harvests energy from the WiFi access point (AP), achieving a power of 30.8 μW at a distance of 30 cm. The output power of the AP was set to 24 dBm. The results state data rates of 5.5 $kbps$ and 2.25 $kbps$ at distances of 30 cm and 40 cm from the AP, respectively. Since then, many systems have been proposed to enhance the key properties of BBCS: consumption, throughput, and range. The trade-off between these three parameters makes it difficult to achieve remarkable performance in all three. Consequently, each contribution aims to enhance the system in concordance with the application. In order to summarize all the ambient BBCS available in the literature, Table 1.3 exposes the vast majority of contributions made to date sorted chronologically. This table summarizes the results of each work as a guide but does not pretend to be a comparison, since it would not be faithful. The lack of standardization in the presentation of results makes it difficult to compare systems since not all the test settings are stated. Maximum communication ranges are highly dependent on the transmitter power and the tag-to-transmitter distance. Throughput and BER can be drastically increased and reduced, respectively, at the cost of increasing consumption or reducing the distance. Carrier modulation affects the system's performance. Unfortunately, the before-mentioned settings are not always exposed in the contributions.

Systems implemented in FPGAs are often accompanied by the design and simulation of an ASIC, yielding results in the range of several tens of μW. On the other hand, systems implemented with commercial off-the-shelf (COTS) components tend to fall within that range for analog modulations but quickly scale up to several hundreds of μW for digital modulations ranging from a few bps to several $Mbps$. The vast majority of proposals present communication speeds in the range of $kbps$. However, some contributions focus on maximizing the speed ((Iyer et al. 2016): 11 $Mbps$, (Liu et al. 2021): 10.61 $Mbps$) at the expense of sacrificing power consumption, a consequence of using higher-frequency oscillators. On the other hand, the communication range varies from 0.3 m (Ishizaki et al. 2011) to 2.2 km (Jiang et al. 2021a). However, this measurement is not very meaningful as it is usually obtained with the tag placed very close to the transmitter. A more reliable characterization would be to evaluate and report the maximum movement area of the tag. LoRa technology-based systems stand out for achieving the highest communication ranges. On the other hand, systems operating in the 2.4 GHz band, such as those based on WiFi and Bluetooth, tend to excel in attaining higher data rates.

1.2.3. *LoRa backscattering systems*

This section provides a more specific review of backscattering systems based on LoRa. Unlike Bluetooth and WiFi-based systems, LoRa signals are not as common nowadays. As a result, LoRa-based systems are often implemented with dedicated transceivers. However, thanks to projects such ashelium or LoRaWAN, the deployment of LoRa gateways has experienced exponential growth in recent years. Helium encourages the installation of LoRa gateways to provide coverage for IoT nodes in smart cities. While most urban areas have a significant number of hotspots, coverage tends to decrease significantly in suburban areas. If the number of LoRa gateways continues to increase, LoRa-based backscattering systems may be able to operate without the need for dedicated transmitters.

A backscattering tag capable of synthesizing LoRa packets using a voltage-controlled oscillator is presented in Talla et al. (2017a). This system uses a frequency tone as a carrier source and a LoRa commercial receiver to demodulate the backscattered signal. A harmonic cancellation mechanism is also introduced to improve the spectral efficiency. The system has been implemented using an FPGA, and its power consumption has been simulated using the Cadence and Synopsis software toolkits. Alternatively, Belo et al. (2019) proposes an IQ impedance modulator to synthesize LoRa packets by modulating the phase of the reflected wave instead of its instantaneous frequency.

Moreover, Peng et al. (2018) presents the implementation of a backscattering tag that uses LoRa ambient transmissions as a carrier source. In contrast to systems based on unmodulated carrier sources, which do not require tag-carrier synchronization, the tag proposed integrates a LoRa packet detection circuit to synchronize the tag modulation with the intermittent LoRa ambient transmissions. The tag achieves that by correlating the incoming signal with a predefined up-chirp saved in its memory. The tag modulates the incident LoRa packets by circularly shifting some of the symbols. An FPGA has been used for the tag implementation. The backscattered signal has been demodulated with a USRP SDR. Experimental results exhibit a maximum data rate of 6.25 $kbps$.

The above-mentioned systems shift the backscattered signal to an adjacent channel to avoid interference of the carrier. The work presented in Guo et al. (2020) modulates the tag information directly in the LoRa carrier band using a simple on–off keying (OOK) modulation. This work proposes a novel algorithm to split the carrier from the tag signal and allow super low power consumption on the tag as a result of the low frequency required by the in-band OOK modulation. The results state data rates of up to 199.4 $kbps$, considerably higher than prior works (Peng et al. 2018). However, the proposed system requires an SDR to demodulate the tag information.

A LoRa backscattering system with parallel decoding, which enables 101 parallel tag transmission, is presented in Jiang et al. (2021a). The tag integrates an envelope detector to detect the presence of LoRa packets. The maximum stated range is 2.2 km under specific conditions (bandwidth: 500 kHz, transmitter power: 30 dBm, tag close to transmitter); however, the wake-up envelope detector circuit limits the range of the tag to 104 m away from the transmitter. Once more, all the implementation relies on an SDR for the demodulation.

Prior works implement a LoRa packet detection by means of a cross-correlation (Peng et al. 2018) and a moving average filter (Guo et al. 2020). However, any of them can demodulate the LoRa packet payload. The work carried out in Guo et al. (2022) copes with this problem and outperforms the detection range with a maximum detection range of 148.6 m in comparison to the 30.6 m and 42.4 m stated in Peng et al. (2018) and Guo et al. (2020), respectively.

A full-duplex LoRa backscattering reader (monostatic architecture) is presented in Katanbaf et al. (2021). The proposed implementation achieves 78 dB of self-interference cancellation in a low-cost, long-range and small-form-factor reader design, making it suitable for integration into smartphones. The reader can demodulate LoRa backscattered packets at distances up to 91.5 m and 15.3 m when the transmission power is set to 30 and 20 dBm, respectively.

Recent research focuses on improving the performance of LoRa backscattering systems. However, many of them achieve it using innovative modulation systems and demodulation algorithms implemented on general-purpose hardware, such as FPGAs and SDRs. Nonetheless, it is challenging to find proposals implemented solely with commercial LoRa transceivers. Despite the recent advances in the field, very few applications can be found leveraging the outstanding properties of LoRa backscattering communications. An automatic irrigation system based on LoRa backscattering is presented in Konate et al. (2020). In Katanbaf et al. (2021), the authors introduce two brief proofs of concept to design a smart contact lens and an aerial surveillance system for precision agriculture. However, the proposed applications are barely developed. Moreover, a long-range cyclist sensing based on LoRa backscattering is presented in Jiang et al. (2021b).

2

Theory of Backscatter Communication

This chapter focuses on addressing the entire theoretical part of the backscattering technique discussed in this book. Thus, section 2.1 begins with a brief introduction to radar cross-section. Section 2.2 deals with the fundamental theory of antennas relevant in the field of backscattering. Finally, sections 2.3 and 2.4 address the differential radar cross-section of the tag and mathematically justify the behavior of the backscattered signal due to the load modulation.

2.1. Radar cross-section

Objects naturally reflect impinging electromagnetic fields based on their radar cross-section (σ), which depends on the object's morphology, including its shape and materials. A larger σ indicates that the target reflects more power back to the source, while a smaller σ means that the target reflects less power back. The σ determines the magnitude of the backscattered power in comparison to that of an ideal sphere with a cross-sectional area of 1 m^2. It should be noted that the radar cross-section also depends on the incident angle. While σ of simple geometric objects can be calculated theoretically, more complex targets require a direct measurement or a simulation. Figure 2.1 shows the maximum radar cross-section (σ_{max}) for three different targets (sphere, flat plate and corner) along with an approximate 180° axial plane pattern to illustrate the differences between them (Center 1997).

2.2. Tag backscattered power

The concept of radar cross-section extends beyond the radar field and plays an important role in any backscattering system. In contrast to the static σ of radar targets (what can be considered a kind of intrinsic analog modulation), backscatter transponders modulate their characteristic σ to encode digital information in the scattered signal. To achieve this, the reflection coefficient of the transponder antenna is modulated by loading it with different impedances. This technique is known as load modulation and is the basis of the RFID and backscattering systems.

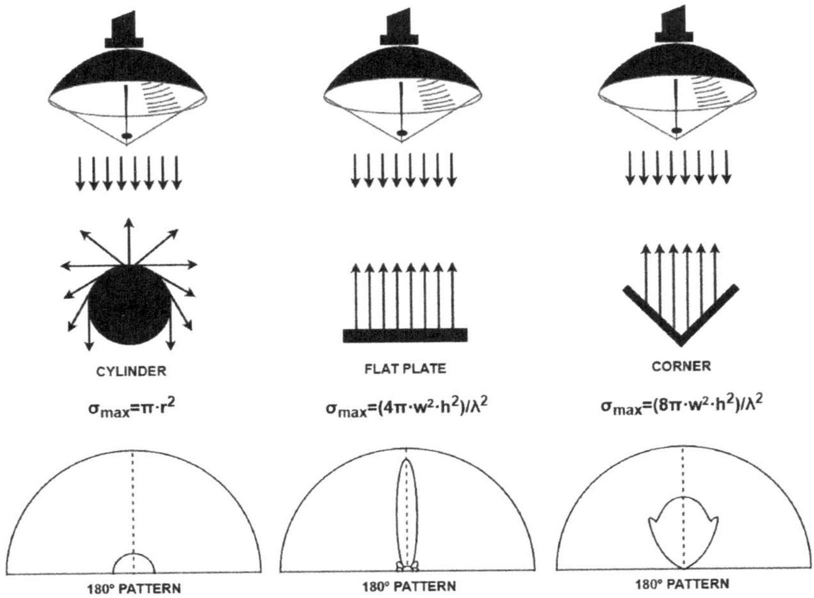

Figure 2.1. *Radar cross-section representation for three different objects: sphere, flat plate and corner*

The theoretical basis behind antenna scattering has been extensively studied in the literature by Harrington (1962), Green (1963), Hansen (1989), Kraus et al. (2006), and more recently by Nikitin and Rao (2006), Yen et al. (2007), Bletsas et al. (2010) and Balanis (2016), among others. Typically, a combination of circuit and field theory is employed to determine the scattered power of a loaded antenna.

The backscattered field can be divided into two main contribution modes called the structural mode and the antenna mode. The structural mode occurs due to the current induced on the surface of the antenna by the interaction with an electromagnetic field. This mode is independent of the load impedance and can be measured when the antenna is conjugate-matched, meaning that maximum power is transferred to the load, leaving only the structural scattering field. On the other hand, the antenna mode depends on the antenna load and applies to any load other than the antenna complex conjugated. Therefore, part of the energy is absorbed by the load, while another part is re-radiated by the antenna due to the reflection caused by the impedance mismatch.

The equivalent circuit of a loaded antenna (or a tag) is depicted in Figure 2.2, where the antenna has been substituted with a Thevenin generator. This method evaluates the antenna mode rather than the structural mode and it is considered an accurate approach for modeling canonical minimum scattering (CMS)[1] antennas (Nikitin and Rao 2006), which are widely used in the UHF band. Despite the inherent limitations of this approach, it can effectively determine the load-dependent power backscattered by the tag (Nikitin and Rao 2006).

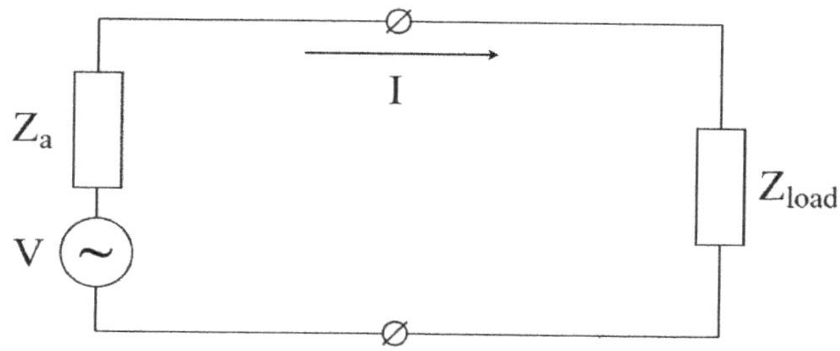

Figure 2.2. *Equivalent circuit of a loaded antenna*

In Figure 2.2, Z_a is the impedance of the antenna ($Z_a = R_a + jX_a$), Z_{load} is the impedance of the load ($Z_{load} = R_{load} + jX_{load}$), V is the open-circuit voltage received by the antenna and I is the current induced by the incident electromagnetic field. In practice, not all power dissipated in the resistance R_a is backscattered; instead, some of it is dissipated as heat in the antenna ($R_a = R_{scatt} + R_{loss}$). However, assuming a lossless antenna ($R_{loss} = 0$), the backscattered power P_a can be mathematically defined as follows:

$$P_a = \frac{1}{2}|I|^2 R_a = \frac{|V|^2 R_a}{2|Z_a + Z_{load}|^2} \qquad [2.1]$$

Likewise, the power delivered to the antenna termination (P_{load}) can also be calculated. When the load impedance is not conjugate matched to that of the antenna, the ratio of power delivered to the load or backscattered will depend on the reflection coefficient at the antenna terminals, mathematically described as:

$$\Gamma = \frac{Z_{load} - Z_a^*}{Z_{load} + Z_a} \qquad [2.2]$$

The Thevenin equivalent circuit does not consider the contribution of the structural mode. Therefore, the field backscattered by the antenna, represented as $\vec{E}a$, can

1 A canonical minimum scattering antenna is defined as one that becomes "invisible" when the accessible waveguide terminals are open-circuited (Kahn and Kurss 1965).

be alternatively described in a more general notation, applicable to any antenna, as follows:

$$\vec{E}_a(Z_{load}) = \vec{E}_s + \vec{E}_{tag}\Gamma \qquad [2.3]$$

where $\vec{E}_s$ is the structural mode, which is the term independent of the load, and $\vec{E}_{tag}$ is the antenna mode (also known as the tag mode). It can be noted that when the antenna is matched with the complex conjugate impedance, the reflection coefficient is zero ($\Gamma = 0$), remaining only the structural scattering field.

Let us consider an impinging electromagnetic wave that generates a power density at the specific location of the tag. The energy flux density of an electromagnetic field passing through a delimited surface is denoted as the Poynting vector $\vec{S}$ (W/m^2) and mathematically defined by the vector product of the electric $\vec{E}$ (V/m) and the magnetic $\vec{H}$ (A/m) field components ($\vec{S} = \vec{E} \times \vec{H}$). As a result of the product of two time-varying fields, the Poynting vector takes different values at different instants of time. This specific magnitude is known as the instantaneous Poynting vector and is not particularly interesting in practice. Therefore, it is typically integrated over a period of time to obtain an average power density (S).

To calculate the average power density delivered by the carrier source to the tag, an isotropic transmitter that uniformly radiates an electromagnetic field over free space is considered. Power is distributed uniformly over the entire surface of the sphere ($P_t/4\pi r^2$). In practice, isotropic transmitters do not exist and the power density depends on the antenna's radiation pattern ($D(\theta, \phi)$). However, by considering that the antenna's maximum gain points towards the tag's coordinates, the power density at the tag can be mathematically determined by:

$$S = \frac{P_t G_t}{4\pi r^2} \qquad [2.4]$$

where P_t is the transmitted power, G_t is the gain of the transmitter antenna, and r is the distance between the transmitter and the tag. The ratio of the power available at the terminals of the antenna (P_{load}) to the incident power density (S) is an area A_{eff} coined as the effective area or effective aperture of the antenna:

$$A_{eff} = \frac{P_{load}}{S} = \frac{|V|^2 R_{load}}{2S|Z_a + Z_{load}|^2} = \frac{\lambda^2}{4\pi} G_{tag} \qquad [2.5]$$

which can be expressed as a function of wavelength (λ) and the tag's antenna gain (G_{tag}) (Nikitin and Rao 2006; Balanis 2016). Any of the powers dissipated in the antenna can be expressed as an aperture: effective aperture (A_{eff}), loss aperture (A_{loss}), scattering aperture (A_{scatt}), and collecting aperture (A_c), which results from the sum of the previous three ($A_c = A_{scatt} + A_{loss} + A_{eff}$). We can note that radar cross-section and antenna apertures have the same units (m^2). This is because, in fact, the radar cross-section of the antenna σ is nothing more than the scattering aperture A_{scatt} of the antenna.

Having introduced the power density, all the powers dissipated in the antenna $(P_{loss}, P_{scatt}, P_{load})$ and its analogous apertures, we can express the radar cross-section with more commonly used parameters (Nikitin and Rao 2006):

$$\sigma = \frac{(\lambda G_{tag} R_a)^2}{\pi |Z_a + Z_{load}|^2} \qquad [2.6]$$

It can be noted that when the antenna is open-circuited $(Z_{load} = \infty)$, the tag does not reflect anything $(\sigma = 0)$. Otherwise, when the antenna is conjugate matched $(Z_{load} = Z_a^*)$, radar cross-section will be equal to the effective aperture multiplied by the tag antenna gain $(\sigma = A_{eff} G_{tag} = \lambda^2 G_{tag}^2 / 4\pi)$. Finally, when the antenna is terminated with a short-circuit $(Z_{load} = 0)$, the radar cross-section depends on the ratio between the real and the imaginary parts of the antenna impedance, reaching the maximum $(\sigma = \lambda^2 G_{tag}^2 / pi)$ when antenna impedance is purely resistive $(Z_a = R_a)$. The radar cross-section of the tag can be expressed as a function of the reflection coefficient (Green 1963):

$$\sigma = \frac{\lambda^2}{4\pi} G_{tag} |C - \Gamma|^2 \qquad [2.7]$$

where C represents a complex parameter independent of the load and related to the structural scattering mode of the antenna (Yen et al. 2007). For instance, when the tag antenna is a thin dipole $C \simeq 1$ (Balanis 2016). Finally, knowing the radar cross-section, the power at the receiver can be determined by means of the classical radar equation:

$$P_r = \frac{P_t G_t G_r \lambda^2 \sigma}{(4\pi)^3 d_t^2 d_r^2} \qquad [2.8]$$

where G_r is the gain of the receiver antenna, and d_t and d_r are the distances from the transmitter to the tag and the tag to the receiver, respectively.

2.3. Tag differential radar cross-section

RFID and backscattering tags modulate the reflection coefficient at their antenna terminals to encode information in the backscattered signal. This is achieved by loading the antenna with different impedances. While most tags use only two states (two loads) for modulation, it is possible to implement high-order modulations with multi-state load-modulated tags. Figure 2.3 shows the equivalent circuit of a backscattering tag, where Z_0 and Z_1 represent two distinct load impedances. The modulating signal that controls the switch is denoted as $p(t)$, the carrier signal as $c(t)$ and the signal reflected by the tag as $r(t)$.

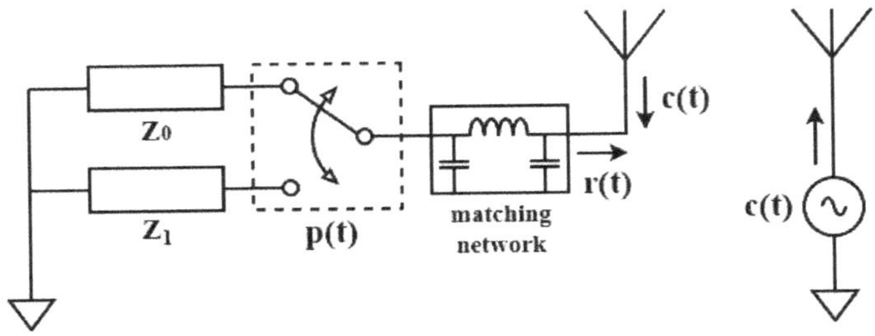

Figure 2.3. *Equivalent circuit of a backscatter tag*

With the switch continuously toggling between two antenna load terminations, the scattering aperture of the tag needs to be redefined based on the reflection coefficients produced by each load. As the information is included in the difference in the backscattered field between the two states, the differential radar cross-section ($\Delta\sigma$) is used. In the context of Figure 2.3, which shows the two impedances Z_0 and Z_1, $\Delta\sigma$ can be mathematically defined as (Nikitin et al. 2007):

$$\Delta\sigma = \frac{\lambda^2}{4\pi} G_{tag}^2 |\Gamma_0 - \Gamma_1|^2 m \qquad [2.9]$$

where Γ_0 and Γ_1 are the complex power wave reflection coefficients associated with Z_0 and Z_1, respectively, and m is the modulation factor that is a function of the first Fourier coefficient of the modulating signal $p(t)$. Assuming a square wave with an ideal 50% duty cycle, m takes a value of $1/\pi^2$. Subsequently, the power at the receiver defined in [2.8] can be recalculated as a function of $\Delta\sigma$:

$$P_r = \frac{P_t G_t G_r \lambda^2 \Delta\sigma}{(4\pi)^3 d_t^2 d_r^2} \qquad [2.10]$$

It can be noted from [2.9] that the power transmitted by the tag depends on the difference between Γ_0 and Γ_1. By using an open-circuit $Z_0 = \infty$ and a short-circuit $Z_1 = 0$, unitary positive and negative reflection coefficients are generated ($\Gamma_0 = 1, \Gamma_1 = -1$), thus maximizing the differential radar cross-section of the tag:

$$\Delta\sigma_{max} = \frac{\lambda^2}{\pi} G_{tag} m \qquad [2.11]$$

Ideally, the two states Γ_0 and Γ_1, are on opposite sides of the real axis of the Smith chart when using an open-circuit and a short-circuit. In practice, however, parasitic elements in the circuit add a reactive component to the load, shifting the reflection coefficients to the complex plane. The key to maximizing $\Delta\sigma$ is to keep the reflection coefficients 180° out-of-phase with each other.

2.4. Load modulation

In its simplest form, using only two loads, the backscattered signal can be modulated in frequency (FSK, Frequency Shift Keying), phase (PSK, Phase Shift Keying) or amplitude (ASK, Amplitude Shift Keying). The phase and amplitude directly depend on the complex impedance observed from the antenna terminals. Specifically, the phase relies on reactance, while the amplitude relies on the resistance of the load. On the other hand, the frequency depends on the load switching frequency.

When the antenna load switches between two states, the result is identical to an RF mixing process, where the incoming RF signal at the antenna $c(t)$ is multiplied by the load switching frequency $p(t)$. From Figure 2.3, assuming that $p(t)$ is close to an ideal square wave signal, inherent in the load-switching process, the resulting backscattered signal $r(t)$ can be described mathematically as follows:

$$\begin{aligned} r(t) &= p(t) \cdot c(t) \\ &= A \cdot cos(\omega_c t + \phi_0) \cdot sgn(cos(\omega_p t)) \\ &= \frac{2A}{\pi} \sum_{n=1,3..}^{\infty} \frac{1}{n} [cos((\omega_c - n\omega_p)t) - cos((\omega_c + n\omega_p)t)] \end{aligned} \qquad [2.12]$$

where A is the amplitude of the backscattered signal, ω_c is the center frequency of the carrier, ω_p is the switching frequency of $p(t)$, ϕ_0 is the phase shift of $c(t)$ relative to $p(t)$ and sgn is the sign function. To ease the mathematical modeling, $c(t)$ has been considered a frequency tone. $p(t)$ is the square wave, synthesized in the load-switching process.

Similarly, we can model the periodically switched reflection coefficient with a square wave of amplitude $\Delta\sigma$ and frequency f_p. Therefore, after the Fourier series development of the square wave, the resulting spectrum, which is given as an infinite sum of delta functions located at each of the harmonics, can be mathematically defined as:

$$\Gamma(f) = \sum_{n=-\infty}^{n=+\infty} c_n \delta(f - (f_c + nf_p)) + \sum_{n=-\infty}^{n=+\infty} c_n \delta(f + (f_c + nf_p)) \qquad [2.13]$$

where c_n are the Fourier coefficients, and f_c is the central frequency of the incident signal at the tag location. For a square waveform with the duty cycle δ_c, the coefficients c_n are given by:

$$c_n = \begin{cases} \frac{1}{2}\Gamma_{avg} & , n = 0 \\ \frac{1}{2}|\Delta\Gamma|\delta_c(\frac{sin(n\pi\delta_c)}{n\pi\delta_c}) & , n \neq 0 \end{cases} \qquad [2.14]$$

where Γ_{avg} is the average power reflection coefficient between the Z_0 and Z_1 states. Using equations [2.13] and [2.14], the backscattered field can be written as:

$$\begin{aligned}\vec{E}_{scatt}(Z_{load}) = (\vec{E}_s + \frac{1}{2}\vec{E}_m\Gamma_{avg})(\delta(f - f_c) + \delta(f - f_c)) \\ + \vec{E}_m \sum_{n\neq 0} c_n \delta(f - (f_c + nf_p) \\ + \vec{E}_m \sum_{n\neq 0} c_n \delta(f + (f_c + nf_p))\end{aligned} \quad [2.15]$$

Therefore, the first term in [2.15] results in the non-modulated term and depends on the structural mode of the tag. The second term represents the modulated sidebands that are a function of the antenna mode. Coefficients of higher amplitude in the Fourier expansion ($n = \pm 1$) are those that result in the components at the frequencies $f_c \pm f_p$.

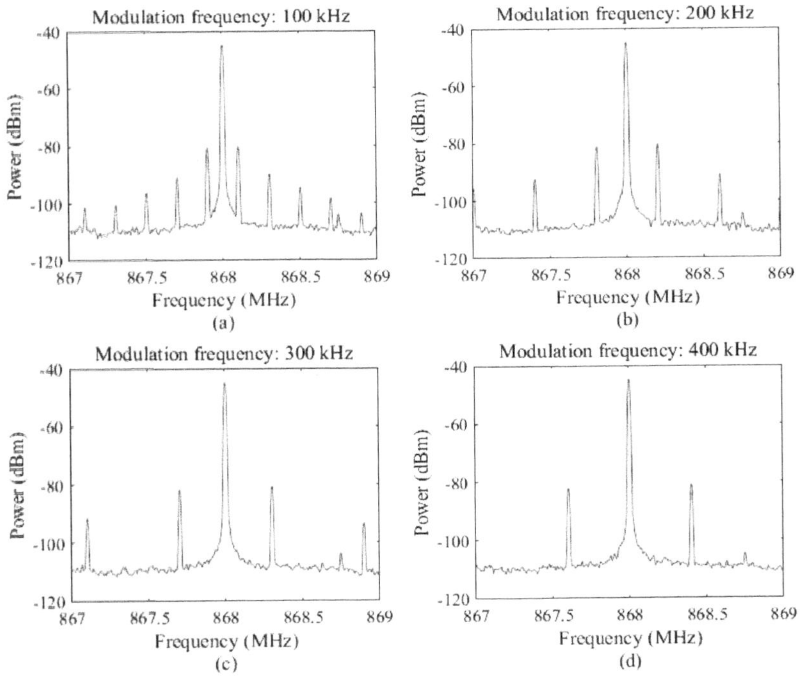

Figure 2.4. *Spectrum of the backscattered signal for four different switching frequencies: 100 kHz (a), 200 kHz (b), 300 kHz (c), and 400 kHz (d). Carrier wave: frequency tone at $f_c = 868$ MHz*

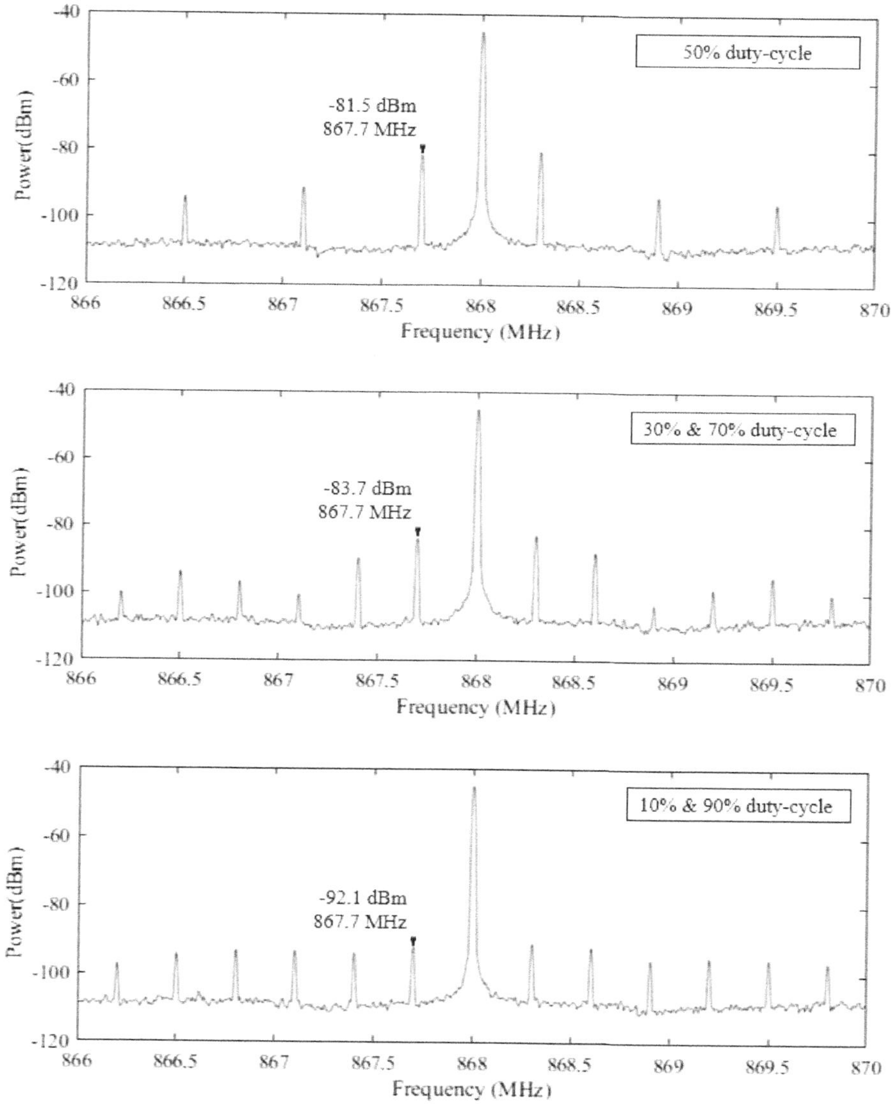

Figure 2.5. *Spectrum of the backscattered signal when the tag is modulated with a square wave of frequency $f_p = 300\ kHz$ for several duty cycles. Carrier wave: frequency tone at $f_c = 868\ MHz$*

Two important drawbacks can be seen from equations [2.12] and [2.15]. First, the load switching generates a double sideband (DSB) modulation with two mirror copies on each side of the carrier ($\omega_c \pm \omega_p$), the backscattered signal and the image. Second, the square-wave signal generates harmonics. To graphically show the two mentioned problems, a frequency tone at 868 MHz has been backscattered. Figures 2.4 and 2.5 show the measurements of the backscattered signal for four different frequencies (100–400 kHz) and for several duty cycles (10%–90%). It can be noted that when the tag is modulated with a 50% duty cycle square signal, the backscattered signal generates half of the harmonics than for other duty cycles (only odd harmonics) and provides the maximum power at these harmonics, as expected from the mathematical development of the Fourier series for a square signal with different duty cycles. Note that the backscattered field at the first harmonic is proportional to the radar cross-section defined in equation [2.9].

3

Backscattering Tag Design

This chapter covers the key elements required to design a backscatter tag and is organized into three sections. The first section examines two front-end implementations: one based on conventional load modulation and another using a reflection amplifier to boost the backscattered signal power. The former minimizes the tag's energy consumption, while the latter trades higher energy consumption for greater communication range. The choice between the two depends on the application requirements. The second section explores two LoRa-based modulation schemes to implement the tag-to-receiver communication. Finally, the third section summarizes the power consumption of the proposed designs and proposes an energy harvesting system to power the tag.

3.1. Tag front-end

3.1.1. *Load-modulated backscattering tag*

Load-modulated backscattering tags can be classified as passive or active depending on the topology implemented in the front-end. While passive tags aim to ensure low-profile and battery-free designs, active tags sacrifice these properties to enhance the communication range. Passive tags integrate an energy harvesting system to power up the tag, typically using RF energy harvesting techniques that take advantage of the ever-present carrier as an energy source. To accomplish this, a portion of the incident RF signal is directed to a rectifying circuit, while the remaining portion is backscattered. Passive tags switch between a matched impedance, ensuring maximum power transfer to the rectifying circuit (harvesting state, $\Gamma_0 \approx 0$), and an

For a color version of all the figures in this chapter, see www.iste.co.uk/lazaromarti/low-power.zip.

unmatched impedance (reflecting state, $\Gamma_1 \leq 1$). This results in a differential reflection coefficient of $|\Delta\Gamma| = |\Gamma_1 - \Gamma_0| \leq 1$. On the other hand, active tags, which lack RF energy harvesting capabilities, can maximize the backscattered power by using antipodal reflection coefficients (ideally, $\Gamma_0 \simeq -1, \Gamma_1 \simeq 1$), resulting in $|\Delta\Gamma| \simeq 2$. However, no load-modulated tag can exhibit a $|\Delta\Gamma|$ higher than 2. Figure 3.1 shows the ideal reflection coefficients for the two aforementioned load-modulated systems.

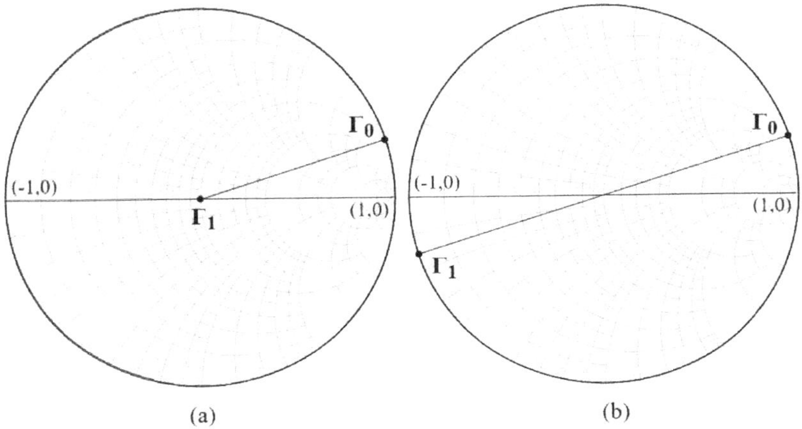

Figure 3.1. *Tag reflection coefficients for two different load-modulation configuration: (a) to maximize $\Delta\sigma$; and (b) integrating energy harvesting capabilities*

Since the passive tags powered by the carrier signal energy are limited by the maximum operating range of the RF energy harvesting circuit, it is better to opt for the implementation of an active tag, allowing us to maximize $|\Delta\Gamma|$, as described in Figure 3.1(a). The system diagram of the implemented load-modulated front-end is the same as the one proposed in Figure 2.3 of section 2.3, when antenna termination is toggled between two states with an RF switch. The two antipodal states have been generated using a short-circuit and an open-circuit. The switch used is an SPDT ADG902 CMOS reflective switch from analog devices with a typical insertion loss of 0.8 dB at 1 GHz (Analog Devices 2003).

A PCB with the ADG902 switch connected to an SMA connector through a 50 Ω transmission line has been manufactured to characterize the front-end with a vector network analyzer (VNA). Figure 3.2 shows an image of the PCB with the ADG902 and a diagram of the measurement setup.

Figure 3.3 shows the measured reflection coefficient when the switch is on the state Z_{SC} (loaded with a short-circuit) and the state Z_{OC} (loaded with an open-circuit). A marker at 868 MHz shows that the two states have a phase difference close to 180°. Losses of 1.5 dB with respect to the ideal short-circuit and open-circuit are obtained due to the insertion loss of the switch.

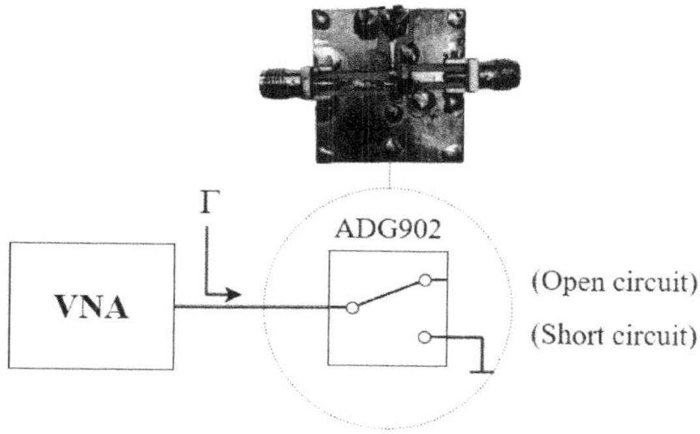

Figure 3.2. *Photograph of the ADG902 characterization board with a diagram of the measurement setup*

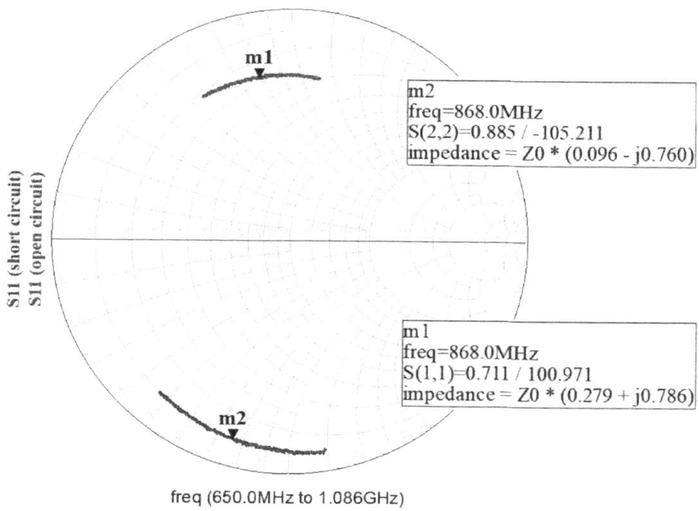

Figure 3.3. *Measured reflection coefficients of the ADG902 SPDT switch for state 1 (open-circuit) and state 2 (short-circuit) at 868 MHz*

Figure 3.4 shows the ratio between the measured differential reflection coefficient and the maximum differential reflection coefficient for an ideal open-circuit and short-circuit ($|\Delta\Gamma|^2/4$). The ratio decreases as the frequency increases due to the higher insertion loss of the switch. The implemented prototype exhibits a loss of -2 dB compared to the ideal case for the 868 MHz ISM band.

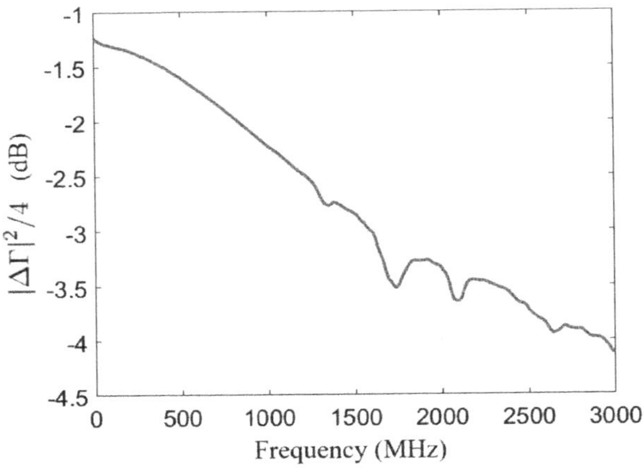

Figure 3.4. *Ratio between the measured differential reflection coefficient and the ideal differential reflection coefficient as a function of the frequency for the ADG902 switch*

3.1.2. *Amplified backscattering tag*

To overcome the differential reflection coefficient constraint discussed in the previous section ($|\Delta\Gamma| \leq 2$), an effective solution is to integrate a reflection amplifier within the tag front-end. To do so, the antenna is directly terminated with the reflection amplifier. When the amplifier is disabled, it generates a reflection coefficient of $\Gamma_0 = Be^{j\phi_0} < 1$, where B and ϕ_0 are the amplitude and the phase of the complex reflection amplifier of the amplifier disabled, respectively. On the other hand, when the amplifier is enabled, it produces a reflection coefficient of $\Gamma_1 = \sqrt{G}e^{j\phi_1} > 1$, where G and ϕ_1 represent the gain and phase of the amplifier in this state, respectively (proportional to the negative impedance generated by the amplifier). By toggling between these two states – enabling and disabling the amplifier – the backscattered signal can be modulated. The maximum value of the differential reflection coefficient, $|\Delta\Gamma|$, occurs when $B = 1$ and $\phi_0 = \phi_1 + \pi$, while the minimum value is obtained when $B = 1$ and $\phi_0 = \phi_1$. Therefore, the differential reflection coefficient falls within the following range: $(\sqrt{G} - 1) \leqslant |\Delta\Gamma| \leqslant (\sqrt{G} + 1)$. However, it is important to note that by turning the amplifier on and off, only one of the two states can take advantage of the amplification. To further increase $|\Delta\Gamma|$, a phase-shift modulator connected between the reflection amplifier and the antenna can be used to generate two antipodal states, while the reflection amplifier is always on, resulting in $|\Delta\Gamma| = 2\sqrt{G}$ (Kimionis et al. 2014a). Figure 3.5 depicts the reflection coefficients when the backscatter signal is modulated, enabling and disabling the amplifier and using a phase-shifter while the amplifier remains always on. This section focuses on implementing the last mentioned system, as depicted in Figure 3.6.

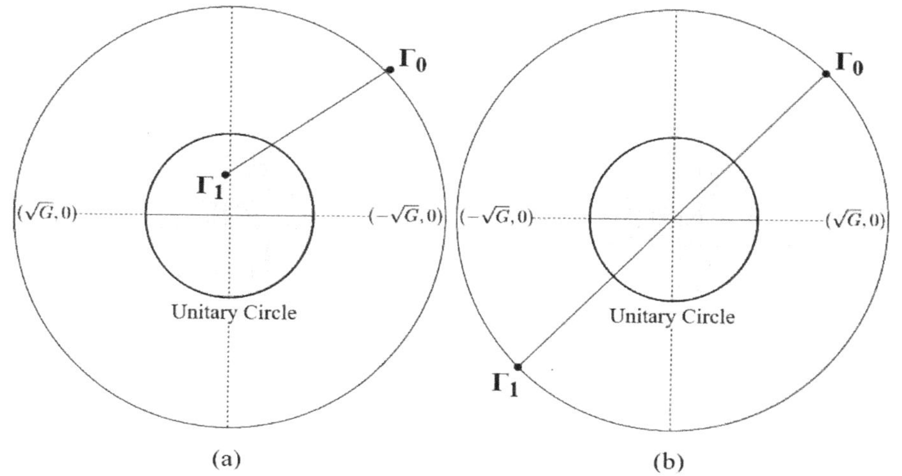

Figure 3.5. *Reflection coefficients of the tag with a reflection amplifier for two different modulation systems: (a) by turning the amplifier on and off and (b) through a phase modulator while the amplifier remains always on*

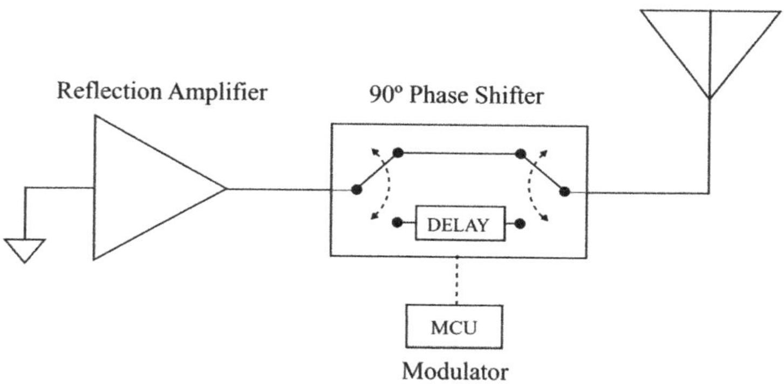

Figure 3.6. *Diagram of the active backscattering tag based on a reflection amplifier along with a phase-shift modulator*

Several works in the literature propose integrating amplifiers into tags to extend the communication range. While some systems suggest using two-antenna topologies, where one antenna receives the carrier signal, amplifies it and transmits it through the other antenna, this approach is not universally applicable. Using two antennas carries the risk of feedback, where the amplified signal loops back through the tag's receiving antenna, resulting in oscillation and interferences in the frequency spectrum. This technique can be used in automotive applications where radar systems operate at

24 GHz (K-band) and 77 GHz (W-band) (Lazaro et al. 2021a). At these frequencies, the isolation between the antennas is high and the signal attenuates rapidly, reducing the likelihood of system oscillation. Moreover, radar applications are suitable for using highly directional antennas, further minimizing the probability of oscillation, as the interaction between pedestrians or vehicles typically occurs in specific directions. On the contrary, in the UHF band, this technique is not commonly used for two reasons: the operating frequency and the use of omnidirectional antennas. At 868 MHz, which is a widely used UHF tag operation frequency, the signal attenuation is much lower, and the use of directional antennas is usually not feasible due to application requirements. Therefore, reflection amplifier-based topologies are more suitable for designing tags in UHF applications, and this section will focus on exploring this topology. Figure 3.7 shows the two different described topologies for the design of amplified backscattering tags.

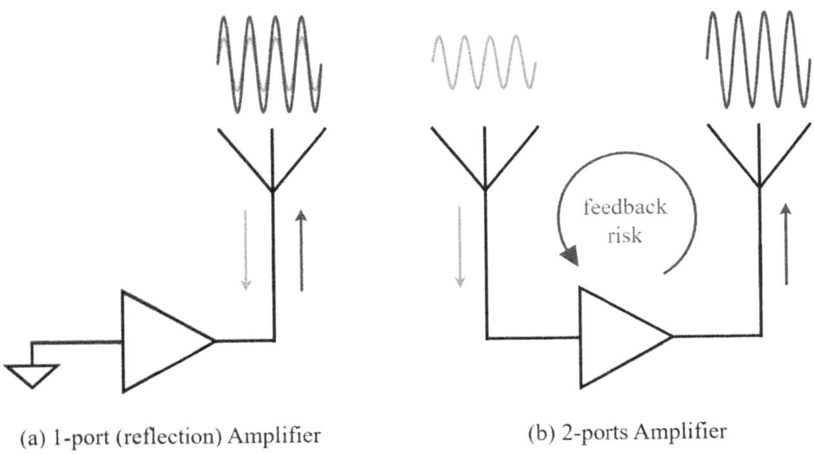

Figure 3.7. *Illustration of a backscattering tag based on a one-port reflection amplifier (a) and two-port reflection amplifier (b)*

Reflection amplifiers are a unique class of amplifiers that harness the principles of wave reflection to achieve amplification. Unlike conventional amplifiers that rely solely on gain elements such as transistors, reflection amplifiers use impedance mismatches and wave interactions to enhance the amplitude of signals. The concept of reflection amplification is rooted in the transmission line theory and the phenomenon of signal reflection. When a signal encounters an impedance mismatch at the interface of two transmission lines or components, a portion of the signal is reflected back toward the source. Reflection amplifiers exploit this reflection by carefully designing the impedance characteristics of the system to achieve signal amplification. By skillfully manipulating impedance mismatches, the reflected signal can be combined with the incoming signal in a constructive manner, resulting in an amplified output (Venguer et al. 2004).

One of the major advantages of reflection amplifiers is their inherent broadband nature, allowing them to operate over a wide frequency range. Furthermore, they can offer high linearity, low noise figures and good power efficiency. These characteristics make reflection amplifiers suitable for applications such as radio frequency (RF) and microwave systems, where high performance and broad frequency coverage are essential (Scanlan and Lim 1964).

Several devices, including IMPATT diodes, tunnel diodes, Gunn diodes, bipolar transistors and FET transistors, can be used in their simplest form to synthesize this type of amplifier. Among them, tunnel diodes have gained widespread attention as an excellent option due to their super-low power consumption (Varshney et al. 2019). However, designs based on bipolar transistors demonstrate comparable performance within the micro-ampere range (Kimionis et al. 2014a). Therefore, both devices will be subsequently assessed and compared for the implementation of low-power reflection amplifiers within the backscattering tag front-end.

A comparison of various previous contributions on the design of reflection amplifiers is provided in Table 3.1. It is worth noting that MBD2057 tunnel diodes exhibit the best efficiency, with an outstanding power consumption of 18 μW. Additionally, using the same device, the maximum gain, with a value of 51 dB is reported for an input power of -100 dBm. Transistor-based models show more modest results, with a maximum gain of 36.19 dB for an input power of -60 dBm. The minimum power consumption reported for transistor-based designs is 325 μW, achieved using a BFT25A bipolar junction transistor.

3.1.2.1. *Tunnel diode-based reflection amplifier*

Tunnel diodes, also known as Esaki diodes, are specialized semiconductor devices that exhibit a unique behavior known as quantum tunneling as a result of the high doping of the p–n junction. Developed in the 1950s by Leo Esaki, tunnel diodes are characterized by their ability to efficiently and rapidly transfer charge carriers through the energy barrier of a semiconductor material. This intriguing phenomenon allows tunnel diodes to operate at extremely high frequencies and to exhibit negative resistance, making them valuable components in various electronic applications. Unlike traditional diodes that rely on majority carrier conduction, tunnel diodes exploit the quantum mechanical phenomenon of electron tunneling. This occurs when electrons can pass through a potential barrier, even though classically they would not possess enough energy to overcome it. The delicate interplay between the band structure of the semiconductor material and the applied voltage leads to a pronounced tunneling effect. Due to their unique characteristics, tunnel diodes find application in areas that require fast switching speeds, high-frequency oscillators, low-noise amplifiers and microwave circuits. Their ability to operate in the terahertz frequency range makes them suitable for cutting-edge technologies such as telecommunications, radar systems and high-speed digital communication (Hall 1960). Therefore, taking advantage of the beforementioned properties, a reflection amplifier based on an AI301A tunnel diode is designed, manufactured and characterized. The circuit schematic for the proposed reflection amplifier is depicted in Figure 3.8.

Ref	Device	Frequency (GHz)	Consumption (μW)	P_{in} (dBm)	Gain (dB)
Matos et al. (2021)	NE3509M04 GaAs HJ-FET	2.37 2.57	1530 5500	-60 -50	36.19 35.83
Hu et al. (2021)	GI307A TD	0.097	90	-50	23
Adeyeye et al. (2021)	MBD2057-E28 TD	5.8	40	-90	50
Gumber et al. (2020)	AI301A GaAs TD	0.41 0.82	144	-43	30 20
Eid et al. (2020)	MBD2057-E28 TD	5.8	18	-100	51
Varshney et al. (2019)	1N3712 Ge TD	0.86	65	-60	35
Khaledian et al. (2019)	NE3509M04 GaAs FET	2.45	-	-	11.5
Amato et al. (2018a)	MBD5057-E28 Ge TD	5.8	45	-75	40
Amato et al. (2018b)	MBD5057-E28 Ge TD	5.8	20.4	-81	35
Farzami et al. (2017a)	NE3509M04 GaAs FET	1.8 (2.4)	2400	-25	21.9 13
Farzami et al. (2017b)	AI301A GaAs TD	0.89	200	-30	17
Khaledian et al. (2017)	AI201A GaAs TD	0.915	178	-30	13
Lee and Yang (2016)	2um Resonant TD	5.8	420	8.6	-25.4
Amato et al. (2015)	MBD5057-E28 Ge TD	5.45	45	-70	34.4
Kimionis et al. (2014a)	BFT25A BJT	0.915	325	-50 -30	10.2 4.9
Kimionis et al. (2014b)	BFT25A BJT	0.9	605	-50 -30	30 14.92
Kimionis et al. (2014c)	BFT25A BJT	0.9	664	-50 -20	29 9
Lee et al. (2014)	InP 0.9 Resonant TD	5.7	125	-	10.4
Chan and Fusco (2013)	BFP405 Si BJT	5.25	2000	-25	13

Table 3.1. *Comparison of the reflection amplifiers proposed in the literature*

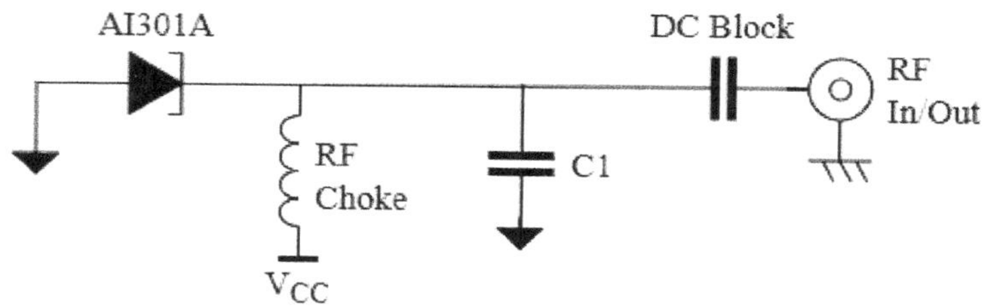

Figure 3.8. *Circuit schematic of the proposed AI301A-based reflection amplifier*

The first step is to characterize the tunnel diode in DC to find the ideal polarization point. DC characterization has been done with a U2722A modular source unit (SMU) from Keysight. Figure 3.9 shows the measured IV polarization curve of the AI301A tunnel diode. At low biasing voltage, tunnel diodes present a differential negative resistance region where current decreases while voltage increases. This characteristic is of particular interest in the design of resonators and amplifiers. As observed from the measured IV curve, the tunnel diode sharply reverses its current at 0.108 mV and 0.64 mV. However, the tunnel diode can only be polarized to leverage the differential negative resistance within the range from 0.108 mV to 0.300 mV, maximizing the gain from 114 mV to 122 mV ($>$ 20 dB).

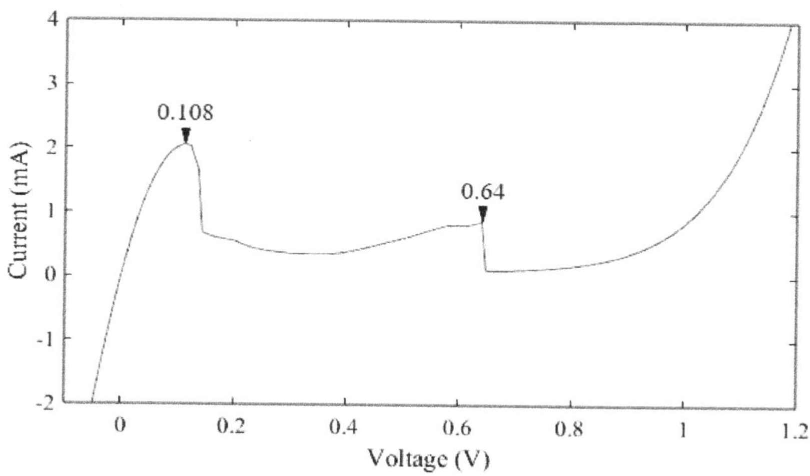

Figure 3.9. *Measured IV curve of the AI301A tunnel diode*

Once the suitable biasing point is determined (114 mV, 1 mA), the S-parameters of the tunnel diode at this biasing point can be measured using the VNA to proceed with the design. The layout of the reflection amplifier has been designed and simulated using ADS. Figure 3.10 shows the layout and a photograph of the final prototype. Capacitor C1 has been connected through a stub to ease the frequency adjustment process. Instead of a straight stub, it has been implemented using meanders to reduce the size of the board.

The reflection amplifier has been manufactured on Rogers 4003C of 32 mils thick. Figures 3.11 and 3.12 show the reflection amplifier gain as a function of the biasing voltage and the input power, respectively.

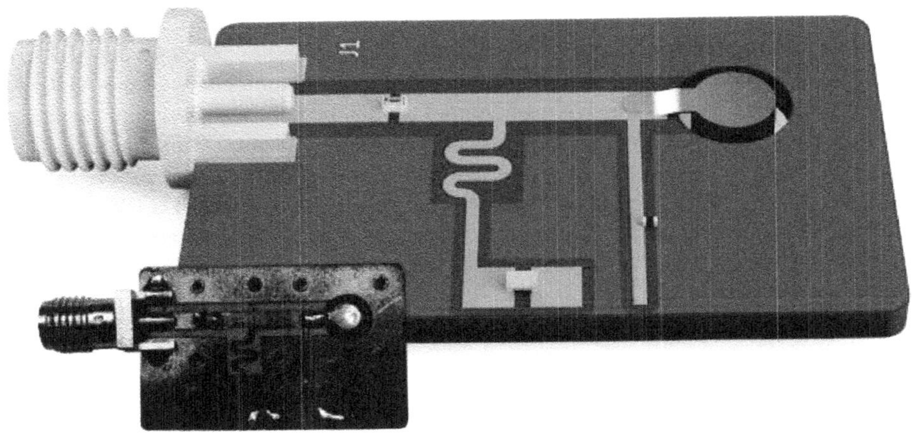

Figure 3.10. *Render and photography of the designed AI301A-based reflection amplifier*

Figure 3.11 shows how the tunnel diode successfully amplifies the reflected signal in the UHF band. It can be observed that the amplifier gain requires a very precise power supply, as variations as low as 3 mV significantly affect the amplifier's behavior.

As expected, in Figure 3.12, we can observe how the amplifier gain decreases with an increase in input power due to the gain compression. In reflection amplifiers, as the input power increases, the reflected signal also becomes stronger, making the active element (the tunnel diode in this case) reach its saturation limit. At this point, the tunnel diode can no longer amplify the signal linearly and starts to distort it.

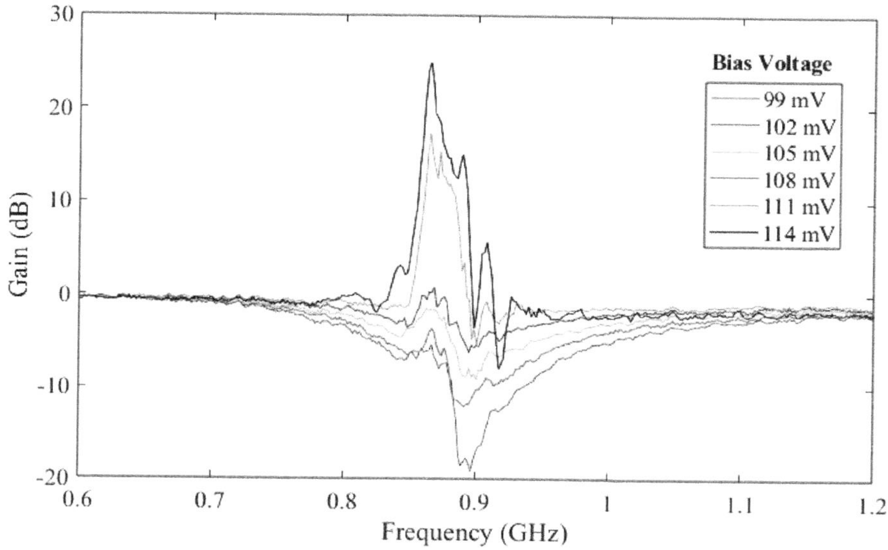

Figure 3.11. *Measured gain (S_{11}) of the Al301A-based reflection amplifier as a function of the biasing voltage*

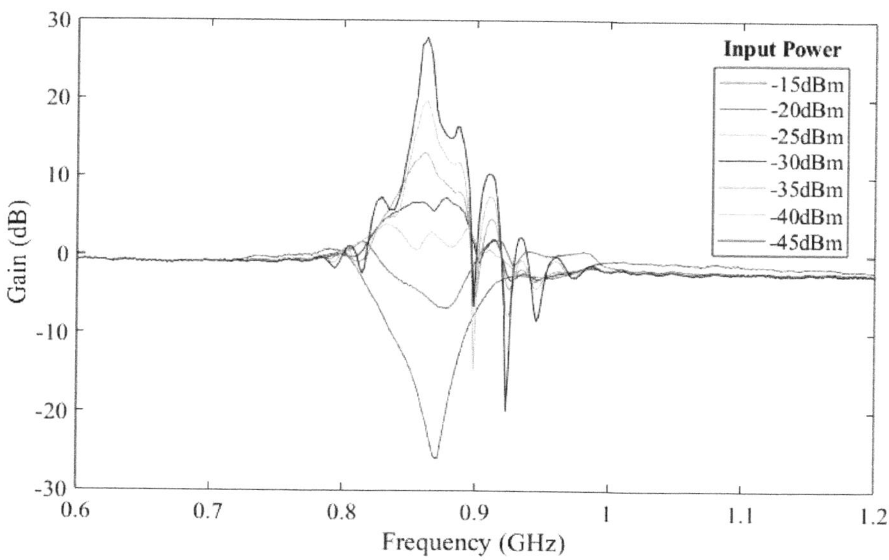

Figure 3.12. *Measured gain (S_{11}) of the Al301A-based reflection amplifier as a function of the input power*

3.1.2.2. *Transistor-based reflection amplifier*

The design of a reflection amplifier with a transistor does not differ much from the design of an oscillator. In fact, the basic principle is the same: to generate positive feedback to amplify the signal at the collector. However, the reflection amplifier is carefully biased to operate near the unstable region, but without reaching oscillation. The design of the reflection amplifier is based on the topology of a Colpitts oscillator. The schematic of the proposed circuit is shown in Figure 3.13.

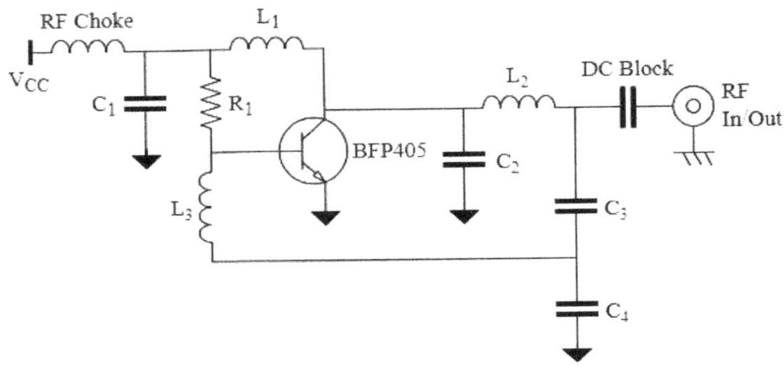

Figure 3.13. *Circuit diagram of the proposed BFP405-based reflection amplifier*

Traditionally, the design of an oscillator or a reflection amplifier involves adjusting the components of the resonant tank circuit to determine the desired oscillation frequency and carefully defines the bias point of the transistor to ensure stable operation. Additionally, impedance matching is essential for optimizing power transfer and overall performance, as it affects the reflected signals fed back to the amplifier, resulting in either an increase or decrease in gain based on their relative phase (constructive or destructive interference).

The reflection amplifier has been designed and tuned in ADS using discrete components. Alternatively, it can be designed solely with microstrip stubs, which reduces costs but increases the layout size. In practice, the use of discrete components simplifies the amplifier adjustment process as it eliminates the need for layout modifications. Hybrid designs are a favorable option as they combine the benefits of both systems, offering cost-effectiveness and flexibility for system adjustments. The layout and photography of the designed and manufactured amplifier are presented in Figure 3.14. The manufactured reflection amplifier has been integrated directly with the phase-shift modulator described in the next section.

The amplifier has been implemented using a BFP405 SIEGET NPN bipolar transistor from Infineon and adjusted to operate at the 868 MHz ISM band. The values of the discrete components used are the following: $C_1 = 470pF$,

$C_2 = 2.8pF$, $C_3 = 6.8pF$, $C_4 = 1.2pF$, $L_1 = 20nH$, $L_2 = 35nH$, $L_3 = 11.6nH$, $R_1 = 355k\Omega$, $C_{DC-block} = 1nF$, $L_{RF-choke} = 100nH$. The biasing point of the reflection amplifier has been adjusted to be 1.8 V, matching the minimum voltage required for any microcontroller.

Figure 3.15 shows the measured IV curve of the amplifier. In contrast to the tunnel diode IV curve, the transistor-based reflection amplifier increases its current continuously while voltage increases. Just like in the previous section, the gain of the amplifier has been characterized based on the bias voltage and input power. The results are depicted in Figures 3.16 and 3.17, respectively.

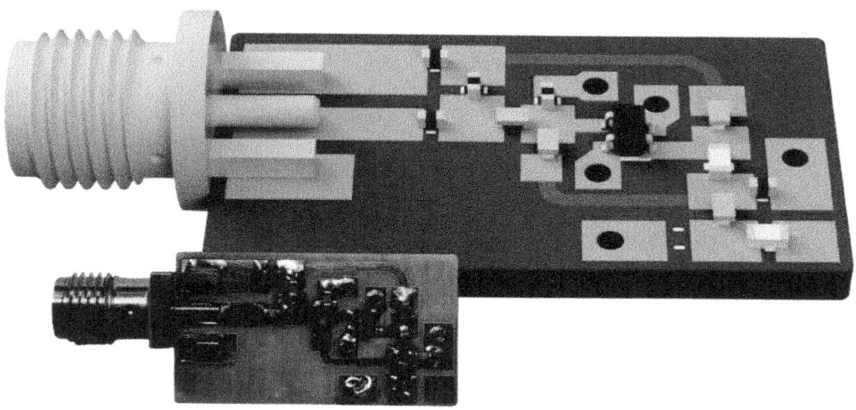

Figure 3.14. *Render and photography of the designed BFP405-based reflection amplifier*

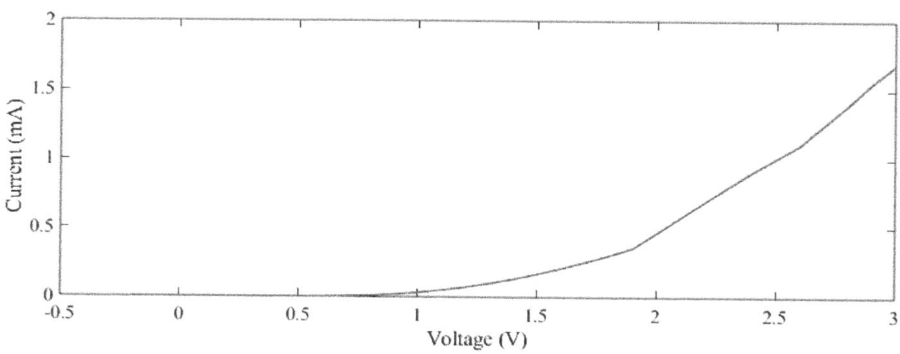

Figure 3.15. *Measured IV curve of the BFP405-based reflection amplifier*

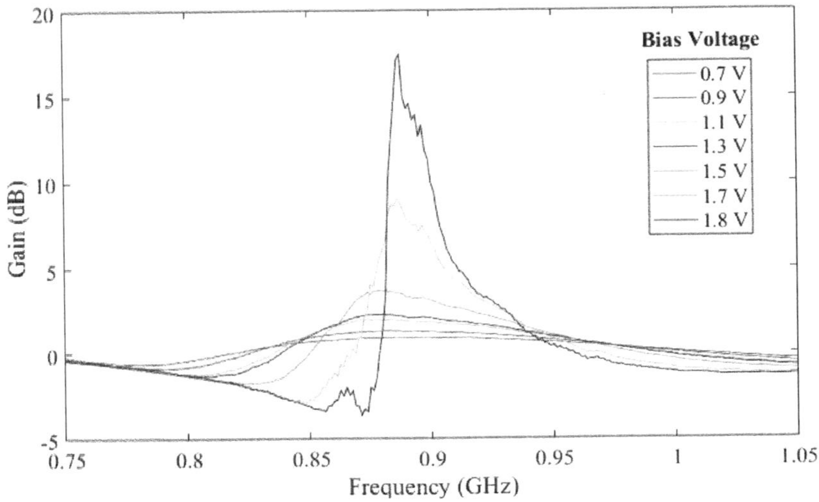

Figure 3.16. *Measured gain (S_{11}) of the BFP405-based reflection amplifier as a function of the biasing voltage*

Although the power consumption of the tunnel diode-based reflection amplifier (114 μW) is significantly lower compared to that of the transistor-based reflection amplifier (612 μW), when the amplifiers have to be integrated within a full system, the tunnel diode version requires us to reduce the minimum supply voltage of the microcontroller by more than one order, from 1.8 V to 114 mV. This down-conversion is not trivial, and no commercially available solutions can be found. A dedicated multi-stage step-down DC–DC converter should be designed with an efficiency high enough to not exceed the power consumption of the transistor version. It seems more favorable to use energy harvesting techniques to power the tunnel diode-based reflection amplifier, but the current consumption and the narrow biasing voltage margin suppose a challenge. A fair comparison between the efficiency of both amplifiers requires considering the power supply conditioning. The transistor-based reflection amplifier has been designed to operate at a voltage of 1.8 V to avoid any conversion losses.

Therefore, although the transistor-based amplifier exhibits a more modest gain-consumption efficiency compared to the tunnel diode when analyzed individually, it proves to be more efficient within the context of a full system integration, since it does not require voltage conversions and has a lower current consumption. Additionally, with the transistor-based design, the biasing voltage can vary almost 1 V, still exhibiting gain.

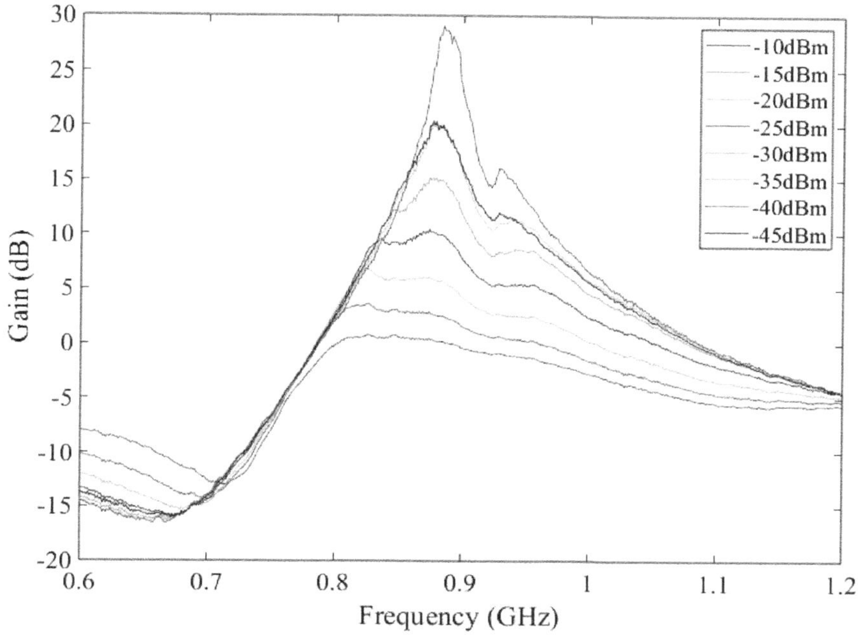

Figure 3.17. *Measured gain (S_{11}) of the BFP405-based reflection amplifier as a function of the input power*

3.1.2.3. *Phase-shift modulator*

As mentioned, the backscattering tag can be modulated by turning the amplifier on and off, thus absorbing or amplifying the impinging signal. However, backscattered power can be maximized keeping the amplifier always on and shifting the phase of the signal by means of a phase modulator. The modulator connects the antenna directly to the reflection amplifier, or alternately through a 90° delay line. As the signal travels twice through the modulator, finally two reflection coefficients shifted 180° one to each other can be generated.

The phase-shift modulator has been implemented using two ADG918 SPDT switches and characterized using a vector network analyzer. Figure 3.18 shows a render and a photograph of the prototype. The size of the phase-shift modulator can be reduced using a meander line instead of a straight line or using a coplanar waveguide instead of a microstrip waveguide to reduce the width of the 50 Ω line.

Figure 3.19(a) and 3.19(b) shows the S-parameter measurements for both states (open-circuit and short-circuit). Figure 3.20(a) and 3.20(b) shows the measurement of the insertion loss and the phase shift for both states overlapped to compare the difference. The insertion loss is close to 1.8 dB in both states, and good matching is observed with a return loss below 24 dB. Modulator insertion loss must be assessed in concordance with the amplifier gain and can be expressed mathematically as (Kimionis et al. 2014a):

$$IL(dB) > -10\log\frac{1}{2}\cdot\frac{1}{2\sqrt{G}} \quad [3.1]$$

where IL is the insertion loss of the phase-shift modulator and G is the gain of the reflection amplifier. If the system design does not ensure compliance with equation [3.1], it is preferable to opt for an on–off amplifier modulation (as depicted at the beginning of the section in Figure 3.5(a)), avoiding the losses inserted by the modulator, and only amplifying one of the two states. However, the measured gain of both reflection amplifiers is high enough to compensate for the round trip insertion loss of the switches (3.6 dB).

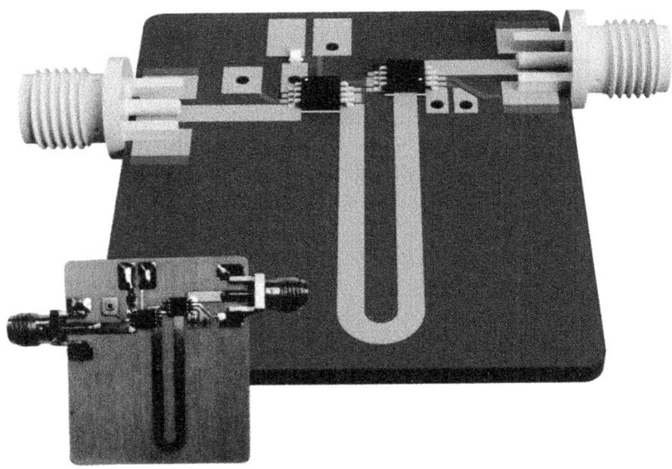

Figure 3.18. *Render and photography of the designed phase-shift modulator*

The phase-shift modulator has been tested with the BFP405 reflection amplifier. Figure 3.21 shows the two states generated by the phase modulator when the amplifier is on (blue) and off (red). It can be appreciated that the blue states are outside the unitary reflection coefficient circle. Since the blue state includes the gain of the amplifier and the losses of the modulator, the reflection coefficient higher than one proves the fulfillment of the equation [3.1].

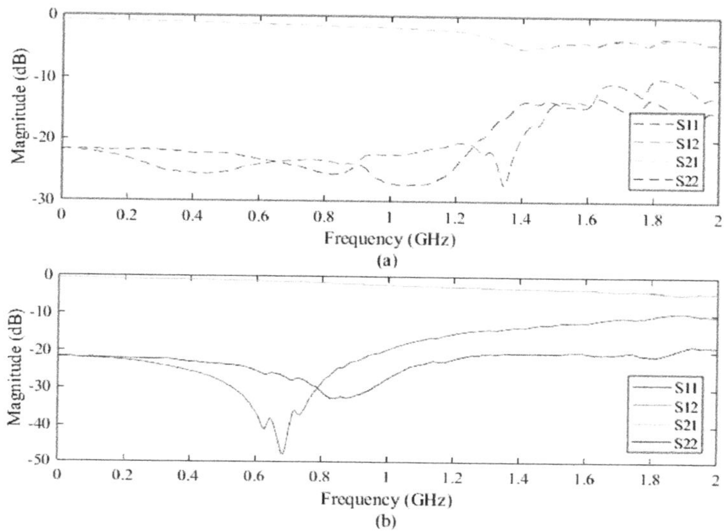

Figure 3.19. *S-parameter measurements of the phase-shift modulator in both states: open-circuit (a) and short-circuit (b)*

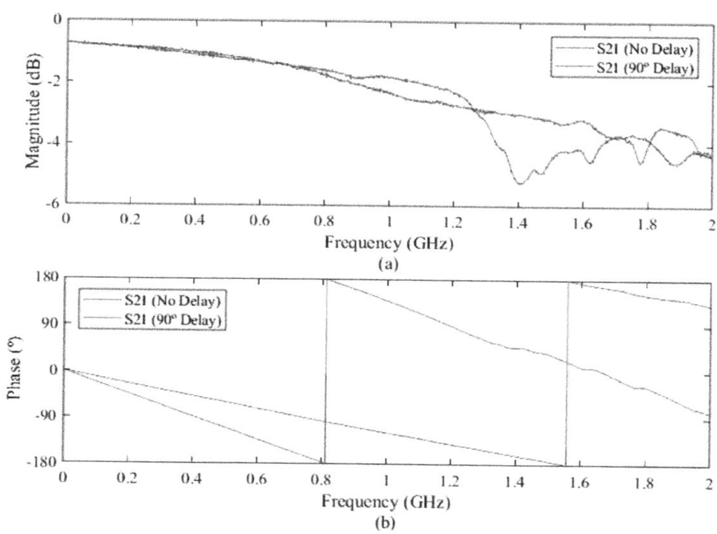

Figure 3.20. *Insertion loss (S_{21}) (a) and phase delay (b) characterization of the reflection amplifier for both states (90º phase shift (blue) and 0º phase shift (orange))*

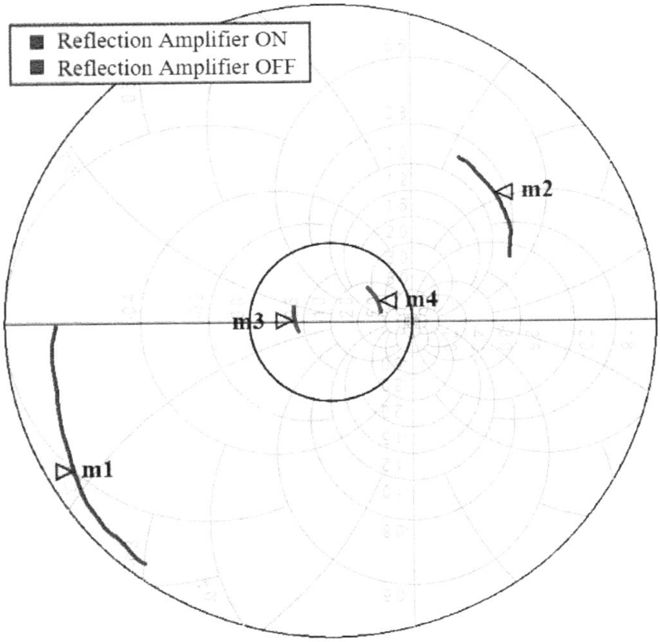

Marker	State	[Mag/Deg]	Impedance
m1	0°	[3.68/ − 149.54°]	$Z_0 \cdot$ (-0.60 − j0.18)
m2	180°	[2.60/39.34°]	$Z_0 \cdot$ (-1.54 − j0.88)
m3	0°	[0.44/177.07°]	$Z_0 \cdot$ (0.39 − j0.02)
m4	180°	[0.63/27.86°]	$Z_0 \cdot$ (2.13 − j2.09)

Figure 3.21. *Measured reflection coefficient S_{11} for the two states generated by the phase modulator (0° and 180° shift) when the amplifier is turned on (blue) and off (red). Frequency range from 855 MHz to 885 MHz. Markers placed at 868 MHz*

3.1.2.4. *Amplifier current monitoring*

Despite the significant research in the literature on the design and integration of reflection amplifiers in the tag's front-end, some challenges need to be addressed further. As observed in the previous sections, the behavior of reflection amplifiers depends on the input power. At certain power levels, there is a risk that the amplifier may start to oscillate, thus generating interference in the spectrum. Furthermore, current consumption increases with higher input powers. Therefore, keeping the reflection amplifier continuously powered in all scenarios does not appear to be a good practice. To address this issue, an active control approach is proposed to enable or disable the reflection amplifier based on its power consumption, which is directly proportional to the incident power. To achieve this, it is suggested to integrate a current sensor into the amplifier's power supply line. A schematic of the system is shown in Figure 3.22.

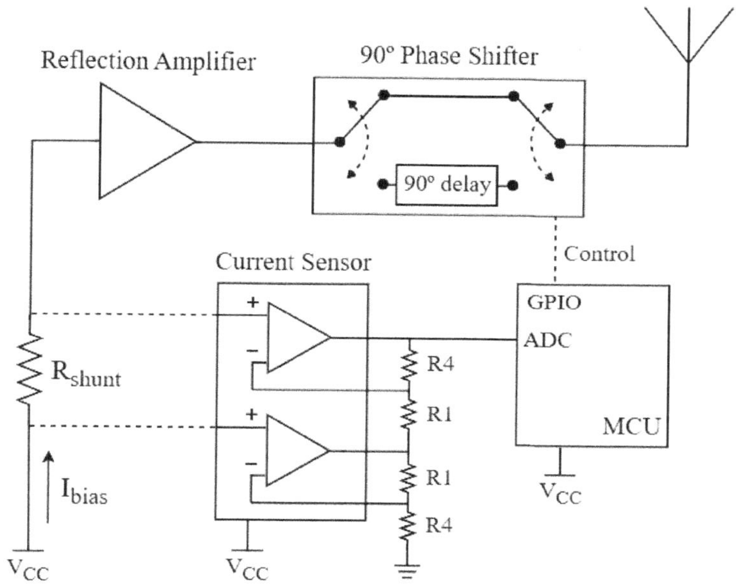

Figure 3.22. *Measured reflection amplifier biasing current as a function of the RF input power*

The current sensor has been implemented by placing a shunt resistance at the input of the reflection amplifier bias. The shunt resistance voltage drop is amplified with an op-amp in differential topology and measured by a microcontroller (MCU). The current sensing accuracy depends on the gain of the amplifier, on the resolution of the MCU analog-to-digital converter (ADC) and on the voltage drop, which in turn depends on the magnitude of the shunt resistance and the current flowing through it.

The current sensor design has been adjusted for an 8-bit Attiny417 MCU, which integrates a 10-bit ADC. Many low-power microcontrollers integrate ADCs of this same resolution. Given a reference voltage of 0.55 V, the ADC can measure voltage variations of 0.537 mV. The reflection amplifier current draw ranges from 280 μA up to almost 1 mA. However, the current sensor is designed to turn the amplifier on and off for current values between 280 μA and 400 μA. Higher currents always disable the amplifier.

A 10 Ω shunt resistance is used, which results in voltage drop from 2.8 mV to 4 mV. Resistors R1 and R4, depicted in Figure 3.22, are set to 1 $k\Omega$ and 100 $k\Omega$, respectively, setting a gain of 101. With this configuration, the operational amplifier output voltage ranges from 282.8 mV to 404 mV. In other words, the minimum detectable current step is 0.5 μA. To implement the current sensor, an MAX9916 has been used, which integrates two 20 μA rail-to-rail Op-Amps. Building the differential amplifier with both op-amps results in a total current draw of 40 μA. However, the current consumption of the sensor can be further reduced to a few μA by using integrated current sensors such as the MAX9634, rendering the power consumption practically negligible compared to that of the reflection amplifier.

Figure 3.23 shows the current consumption of the amplifier in relation to the input RF power. Oscillations have been observed in the spectrum analyzer for input powers exceeding -25 dBm. However, even at -40 dBm, a marginal increase in current can be discerned, providing an opportunity to deactivate the amplifier, thus reducing the current consumption of the system to 300 μA. With power levels of approximately -40 dBm, the backscattering tag can effectively work without the need for the reflection amplifier.

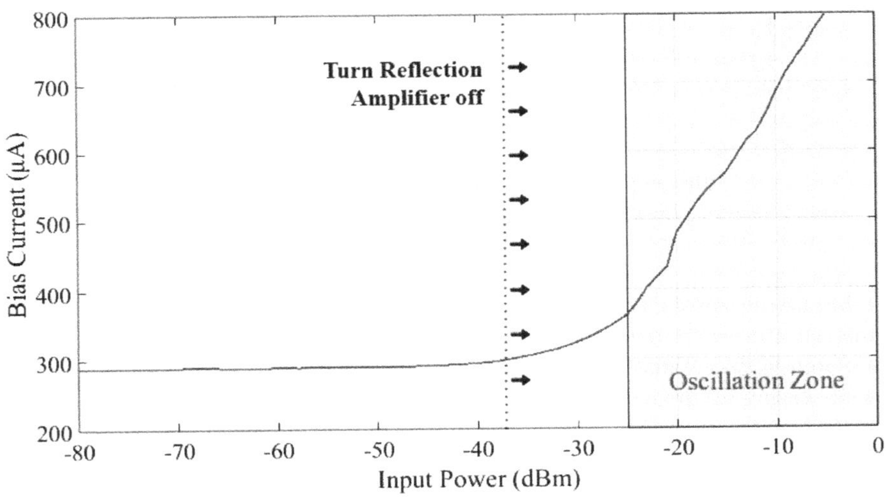

Figure 3.23. *Measured reflection amplifier biasing current as a function of the RF input power*

Although in practice many wireless transmissions tend to be below -40 or -50 dBm at a few meters from the transmitter, this is not always the case. For example, a commercial LoRa transmitter, which can emit 20 dBm, with a monopole antenna (3 dBi gain) can perfectly reach -25 dBm at a distance of 3 m in direct line-of-sight.

Considering that backscattering tags are designed for the IoT field, where many nodes coexist, it is not uncommon to assume that some of them could eventually be located a few meters or centimeters away from the transmitter. In such cases, these nodes may start to oscillate, consequently disabling the communications of all of the other nodes. Thus, the active control of this type of amplifier is mandatory if they are integrated into backscattering tags. This practice becomes even more necessary in reflection amplifiers based on tunnel diodes, which present a less stable behavior.

3.1.3. *Backscattered power comparison*

This brief section shows the gain introduced by the reflection amplifier compared to the conventional load-modulated front-end. Both tunnel diode-based and transistor-based reflection amplifier gains have been compared to the conventional front-end. A measurement has been taken in monostatic configuration with two 9 dBi UHF RFID antennas located 1 m away from the tag. Absorbing panels have been used behind the tag to reduce interferences. Measurements are presented in Figure 3.24. The peak backscattered power in the first harmonic is -95.05 dB, -78.65 dB and -76.55 dB for the conventional load-modulated, the transistor-based, and the tunnel diode-based reflection amplifiers, respectively. The switching frequency of the tag has been set to be 100 kHz. The SIEGET BJT transistor has been polarized at 1.75 V and the tunnel diode at 114 mV.

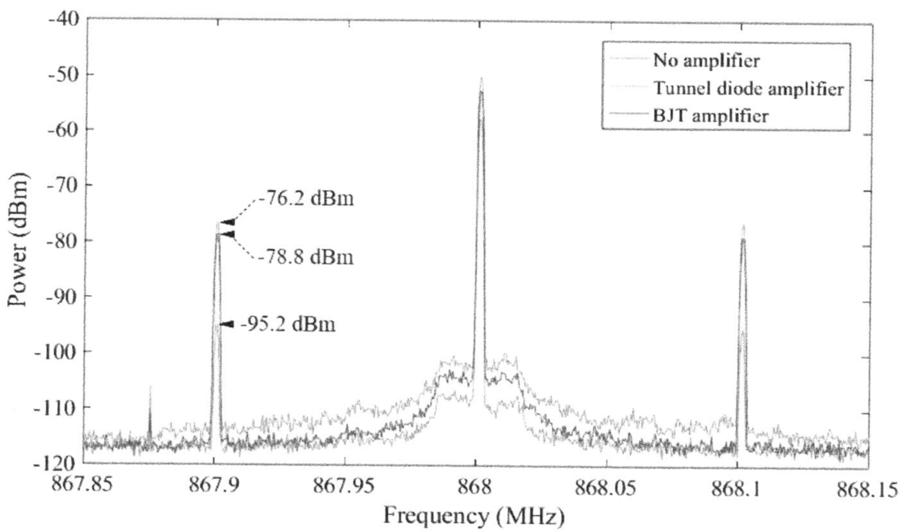

Figure 3.24. *Comparison between the proposed front-ends: passive load-modulation (green), transistor-based reflection amplifier (red) and tunnel diode-based reflection amplifier (blue)*

3.2. Tag and carrier modulation

The backscattering systems proposed in this book use LoRa transmissions as the carrier signal, unlike other works that are based on other types of modulations (WiFi, BLE, etc.) or direct non-modulated carriers (frequency tones). This section discusses the modulation of the carrier and tag. Firstly, an introduction to LoRa modulation is provided in order to understand the advantages it brings when used as a carrier. Then, two different tag modulations are presented: one implemented by simply enabling and disabling the backscattered signal, and another consisting of modulating the tag's switching frequency to generate a down-chirp signal that is inverse to the impinging up-chirps of the carrier.

Before proceeding, it is necessary to clarify the following: in previous sections, the term modulation has been used to refer to the process of switching between multiple loads or turning the amplifier on and off. In this section, we will also address tag modulation, but at a higher level, discussing the architecture of the backscattering system and data transmission methods used in practice.

Throughout this section, the same nomenclature defined in section 2.3 is used, where $p(t)$ represents the signal that controls the switch, $c(t)$ is the carrier signal, $r(t)$ is the backscattered signal, and Z_0 and Z_1 represent the two termination states of the antenna, Z_0 being an open-circuit and Z_1 being a short-circuit.

3.2.1. *Carrier modulation: LoRa primer*

LoRa[1] is a wireless modulation technology patented by Semtech, specially designed for long-range communications with a low power consumption (Semtech Corporation 2015). It stands out for its ability to provide robust connectivity over long distances while operating in unlicensed frequency bands (Noreen et al. 2017). LoRa modulation is specifically designed to optimize the trade-off among range, capacity and power consumption, making it suitable for a wide range of applications, especially for the Internet of Things (IoT). It operates in the sub-GHz frequency bands, typically using frequencies such as 433 MHz, 868 MHz or 915 MHz, depending on the regulatory domain (Asia, Europe or America). Recently, Semtech released a LoRa IC operating at 2.4 GHz (Deréviankine et al. 2023), enabling worldwide operation without being constrained by specific band regulations. The use of these lower frequencies enables LoRa signals to propagate over longer distances and penetrate obstacles more effectively than higher-frequency wireless technologies.

1 LoRa, which stands for long range, defines the physical (PHY) layer of the open systems interconnection (OSI) model. On the other hand, LoRaWAN is a medium access control (MAC) layer protocol maintained by the LoRa Alliance. These two terms should not be confused.

LoRa uses a proprietary modulation scheme called chirp spread spectrum (CSS), which modulates the data onto chirp[2] by circularly shifting them as described in Figure 3.25.

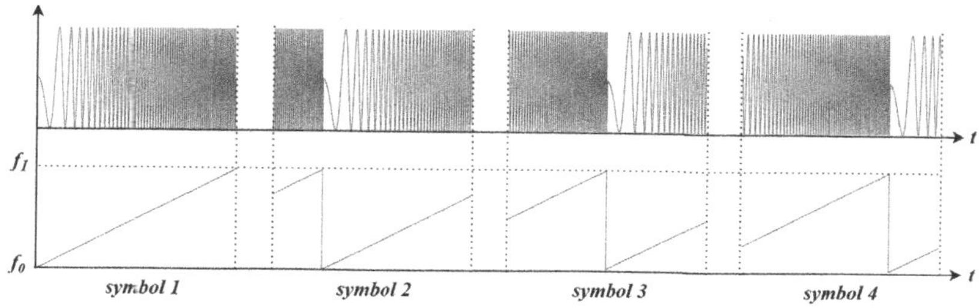

Figure 3.25. *Chirp circularly shift modulation used in LoRa to encode symbols*

Figure 3.26 shows the direct measurement of a LoRa packet. The wide bandwidth of the chirp signals enables LoRa devices to achieve high sensitivity and robustness against noise and interference, allowing LoRa to coexist with other wireless technologies in the same frequency band. Furthermore, the use of error correction techniques ensures reliable communication even in harsh environments with high levels of interference. LoRa transceivers typically allow users to adjust the chirp bandwidth from 7.8 to 500 kHz; however, only three of these (125, 200 and 500 kHz) are supported by the LoRaWAN protocol. LoRa supports different spreading factors (SF), ranging from 7 to 12, which determine the data rate (from 0.3 $kbit/s$ to 27 $kbit/s$) and communication range (up to 5 and 15 km in urban and rural areas, respectively). Higher SF provide a longer range but lower data rates, while lower SF offer higher data rates but a shorter range. This flexibility allows LoRa to adapt to various application requirements, balancing the trade-off between range and data throughput (Semtech Corporation 2015).

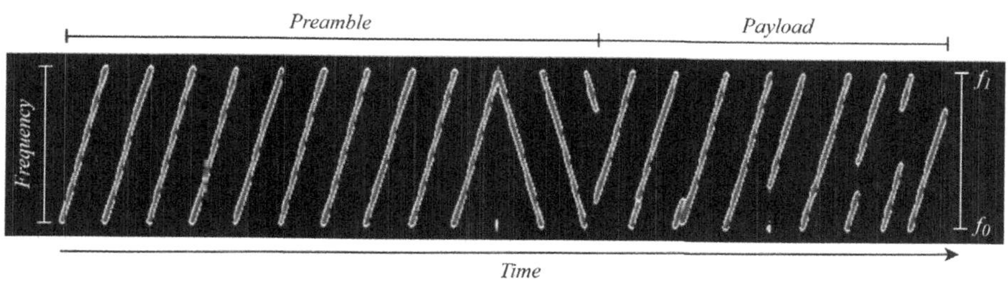

Figure 3.26. *Measurements of a LoRa packet transmission*

2 A chirp, often called a sweep signal, is a signal in which the frequency increases (up-chirp) or decreases (down-chirp) with time.

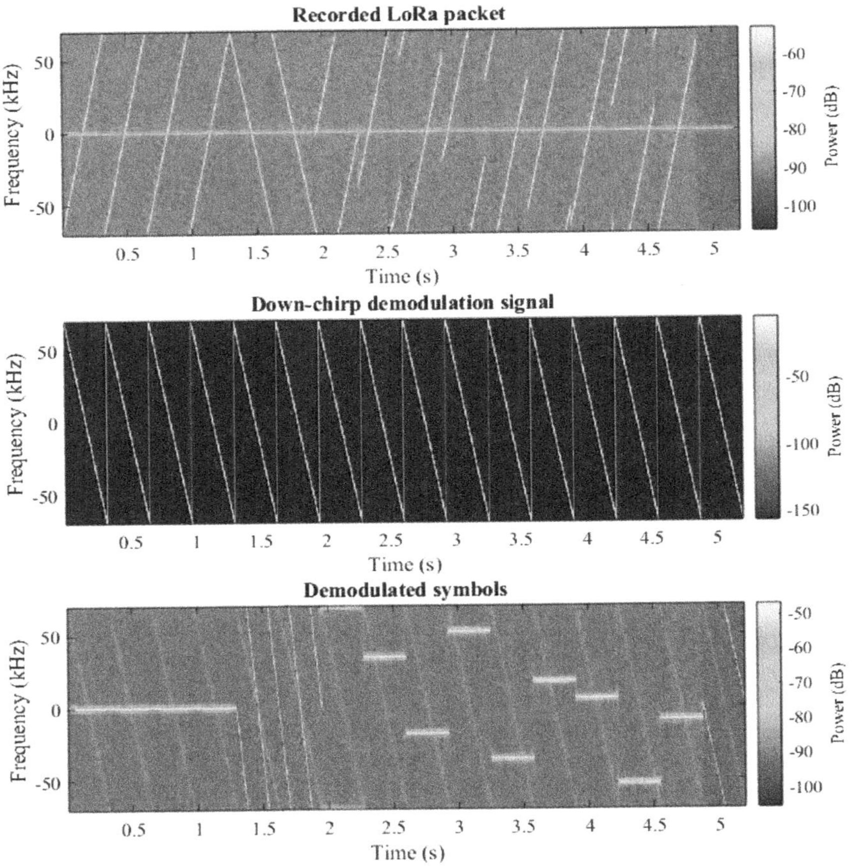

Figure 3.27. *Matlab LoRa packet demodulation*

The remarkable communication range of LoRa modulation is attributed to the high sensitivity of the receivers, which can reach as low as -148 dBm (Semtech Corporation 2015). This sensitivity allows the demodulation of signals up to 20 dB below the noise floor, surpassing the capabilities of other modulations (Thomas et al. 2012). This outstanding property of LoRa modulation is indeed attributed to the demodulation process performed at the receiver, where the incoming up-chirps are multiplied by down-chirps of the same period and bandwidth, resulting in a frequency tone with a high processing gain. Figure 3.27 demonstrates the measurement and demodulation of a LoRa packet using MATLAB. The figure shows the recorded up-chirps of the LoRa signal, the down-chirp demodulation sequence and the resulting demodulated levels obtained through the multiplication of these two.

Therefore, among all of the mentioned advantages of LoRa modulation, the processing gain, receiver sensitivity and kilometer-range communication are key features that make LoRa modulation ideal for compensating the weak backscattered power from the tag. LoRa significantly enhances the communication of the backscattering system, which would be challenging with any other modulation type. Table 3.2 summarizes the characteristics of the main wireless technologies (ZigBee Alliance 2017; Z-Wave Alliance 2017; Sigfox 2019; 3rd Generation Partnership Project 2020a, 2020b; Bluetooth Special Interest Group 2020; Wi-Fi Alliance 2020; Semtech 2021; Thread Group n.d.).

It can be observed that LoRa receivers exhibit the best receiver sensitivity. Other modulations such as NB-IoT, LTE-M and Sigfox also achieve a kilometer-range communication; however, they have some disadvantages compared to LoRa. NB-IoT and LTE-M operate in licensed spectrum bands and require coordination with cellular network operators. Moreover, cellular-based technologies such as NB-IoT and LTE-M typically consume more power than LoRa. On the other hand, Sigfox operates as a subscription-based service with proprietary network architecture and communication protocols. Additionally, Sigfox has a significantly lower data rate, approximately 500 times lower than that of LoRa.

Wireless Technology	Frequency	Data Rate	Modulation	Receiver Sensitivity	Range
WiFi 6 (802.11ax)	2.4 or 5 GHz	Up to 9.6 $Gbps$	OFDMA	-82 dBm	Up to 36 meters (NLoS)
Bluetooth 5.2	2.4 GHz	Up to 50 $Mbps$	GFSK	-90 to -95 dBm	Up to 240 meters (LoS)
ZigBee	2.4 GHz	Up to 250 $kbps$	O-QPSK	-101 to -103 dBm	Up to 100 meters
Z-Wave	800 - 900 MHz	Up to 100 $kbps$	GFSK	-103 to -108 dBm	Up to 30 meters
LoRa	433/868/915 MHz	Up to 50 $kbps$	CSS	-126 to -148 dBm	Up to several kilometers
NB-IoT	Cellular	Up to 250 $kbps$	QPSK	-129 to -134 dBm	Up to several kilometers
LTE-M	Cellular	Up to 1 $Mbps$	QPSK 16QAM	-116 to -123 dBm	Up to several kilometers
Sigfox	863 - 925 MHz	Up to 100 bps	UNB (BPSK)	-126 to -132 dBm	Up to several kilometers
Thread	2.4 GHz	Up to 2 $Mbps$	OQPSK	-100 to -110 dBm	Up to 100 meters

Table 3.2. *Main wireless technologies*

3.2.2. *Tag modulation: on–off keying*

The backscattered signal's amplitude, phase and frequency can be modulated based on the load properties and switching frequency, as described in the first section of this chapter. In contrast to the conventional ASK and FSK modulations commonly used in RFID, the modulation proposed in this book is based on the principle coined as frequency-shift (FS) backscattering (Zhang et al. 2016a). The FS-backscattering technique involves shifting the backscattered signal to a non-overlapping adjacent channel by periodically switching between the loads at a constant frequency. By relocating the backscattered signal to a clean adjacent channel, the interference generated by the carrier is minimized, thus enhancing the signal-to-noise ratio (SNR) at the receiver. Once the frequency shift is applied, the tag's digital information can be encoded using various methods. For instance, by turning the switching signal $p(t)$ on and off, we can implement a binary amplitude modulation in the adjacent channel, commonly known as on–off keying (OOK) modulation. On the other hand, by switching at two different frequencies, a binary frequency-shift keying (B-FSK) modulation can be achieved. It should be noted that the primary distinction between traditional and FS-backscattering modulations lies in the constant frequency switching used to shift the backscattering uplink toward an unoccupied channel. Figure 3.28 shows the difference between the control signal $p(t)$ for the implementation of an ASK modulation using the traditional way or using the FS-backscattering technique.

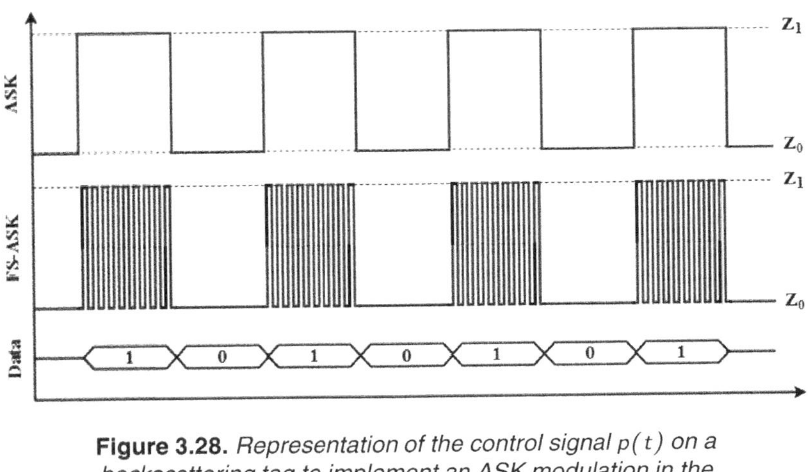

Figure 3.28. *Representation of the control signal p(t) on a backscattering tag to implement an ASK modulation in the traditional way or using the FS-backscattering technique*

The proposed system to implement the tag-to-receiver uplink communication is a LoRa-packet OOK modulation. The system deployment is depicted in Figure 3.29. A LoRa transmitter is set to generate the carrier signal at a center frequency f_c. The tag, which is modulated at a frequency f_p, backscatters the impinging LoRa packets to an adjacent channel ($f_c \pm f_p$) where the LoRa receiver is listening. It can be noted that the receiver cannot directly demodulate the LoRa packets of the transmitter, but only those backscattered by the tag. The advantage of this method is that neither the carrier transmission nor the shifted packets are modified at all, allowing any commercial off-the-shelf LoRa receiver to directly demodulate the backscattered packet. The communication system implements a packet-level modulation, where digital ones and zeros are encoded in the presence or non-presence of packets. This communication system does not alter the original LoRa transmission and is independent of the packet payload.

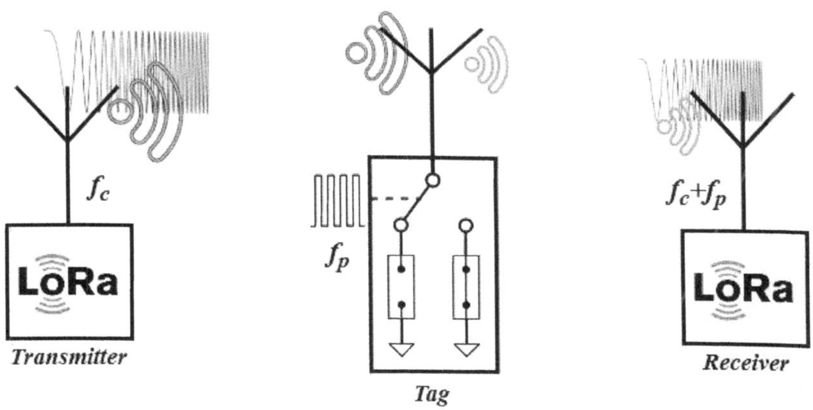

Figure 3.29. *Diagram of the system architecture for the proposed OOK modulation*

The LoRa transmitter node is continuously sending packets, whose duration (packet air-time) depends on the number of symbols sent, which in turn depends on the LoRa packet configuration: SF, bandwidth, preamble length (default to 8), coding rate (CR), sync word (default to 0x12) and, of course, the payload. For example, for a packet with the following configuration: SF: 12, BW: 125 kHz, CR: 1, preamble: 8 symbols, and payload: 10 bytes), the time on air is approximately a second (991.23 ms).

Figure 3.30 shows the spectrogram and waterfall plot obtained from measurements of the OOK modulation described in this section. The spectrogram is generated using the phase and quadrature (IQ) samples captured with a software-defined radio (SDR). The waterfall plot is created by stacking the instantaneous measured power level within a periodic frame of time. The LoRa transmitter is configured to transmit packets at 868 MHz with an SF of 12 and a bandwidth of 125 kHz. The backscattering tag's switching frequency was set to 300 kHz, generating two sideband channels at 867.7 MHz and 868.3 MHz. By disabling the tag's switching frequency, it is possible to observe how the sideband transmissions are cut off.

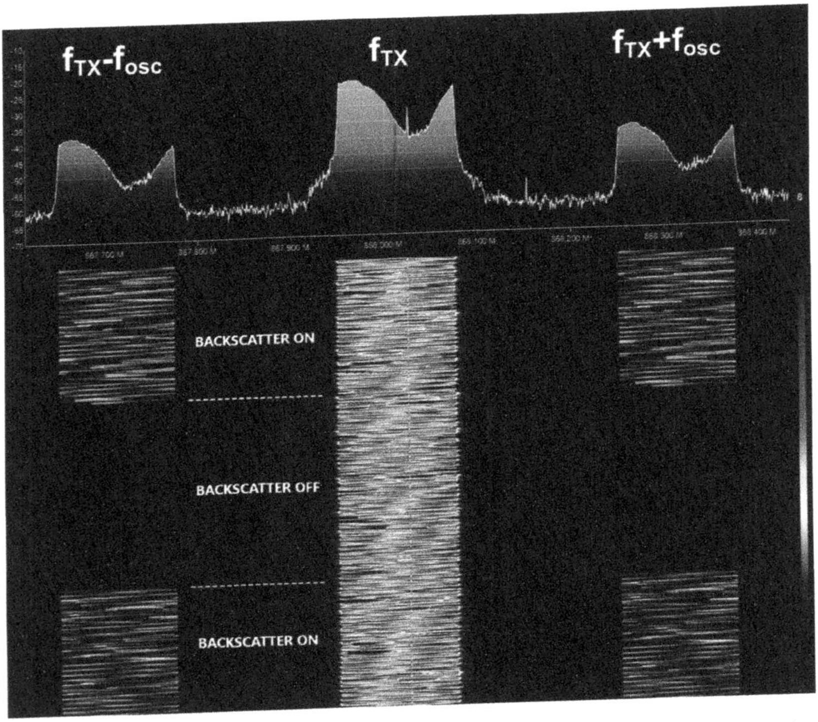

Figure 3.30. *Measured spectrogram with backscatter on and off showing the shift of the LoRa channel to offset channels when the backscatter is on*

The communication has been implemented following the scheme depicted in Figure 3.31, based on the well-known asynchronous serial port protocol. Tag data is encoded in frames. The frame length depends on the number of bits required to send the tag information, which in fact varies with the application. Frame length can be fixed or variable. In this last case, it is recommended to send a field within the frame indicating the total length of the frame. Additionally, parity bits, checksums or CRC bits can be added to enhance transmission reliability. For the proof of concept, a byte is more than enough to encode a temperature value, an identifier or a simple indicator. The tag sends a byte of data and two start/stop bits to indicate the start and the stop of the payload. Therefore, the frame length is 10 bits. The LoRa packet air-time has been adjusted to be the minimum as possible, thus increasing the communication data rate. Therefore, for an SF of 7, a bandwidth of 250 kHz, a preamble length of 6 and the minimum payload, the packet air-time is reduced to 18 ms. The duration of each bit (32 ms) is the sum of a LoRa packet and the gap between each one of them, which has been set to be 14 ms. As the backscatter leaves a minimum time between frames equal to its frame length ($10 \cdot 32\ ms$) and considering the values mentioned above, the system can send frames every 640 ms; therefore, the data rate is 12.5 $bits/s$.

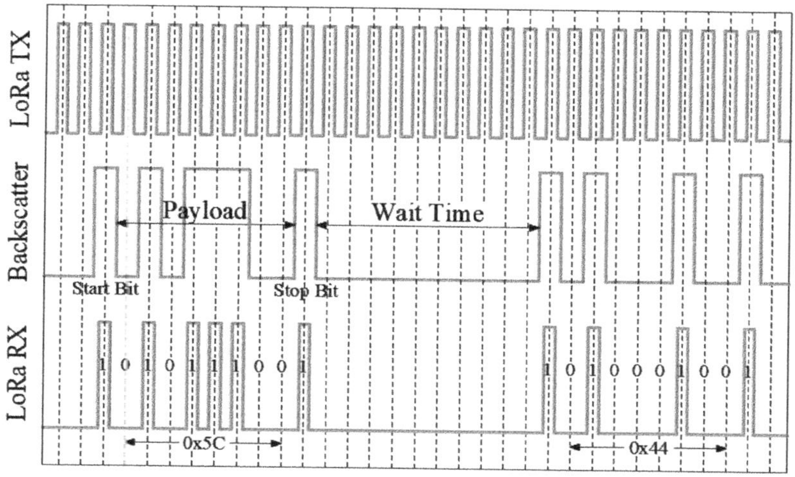

Figure 3.31. *Block diagram showing the data transmission using the reflection of the transmitted packets*

Depending on the scenario, there are several strategies for managing collisions between tag transmissions. One method is to deliberately reduce the communication range of the tags if the application allows it. By increasing or reducing the SF and bandwidth of LoRa packets, the processing gain at the receiver varies proportionally, thereby increasing or reducing the range. This approach effectively reduces collisions since only tags within a constrained area can send packets. Another method involves randomly adjusting the waiting time between frame transmissions, which eliminates periodic collisions between transmissions from different backscatters.

The communication protocol can be implemented in a low-power 8-bit microcontroller (MCU) from the AtTiny family from Microchip. The switch control signal $p(t)$ can be generated in several ways, using the integrated pulse width modulation (PWM) function of the MCU or integrating an oscillator in the tag which is enabled or disabled by the MCU. These two methods have their own advantages and disadvantages. While the dedicated oscillator exhibits a higher-frequency accuracy, it normally consumes more than the PWM function of the MCU. Fortunately, LoRa receivers can measure the frequency shift from the expected central frequency and the packet received central frequency, dynamically correcting the frequency offsets due to the backscatter oscillator drifts, whether it is the internal of the MCU, or the dedicated one. The optimal system should be studied based on the specific components, as running the entire MCU at 300 kHz may be less efficient than running it at 32 kHz to control an external oscillator of 300 kHz. It all depends on the power consumption of the MCU running at 300 kHz with all of the functions disabled (ADC, DAC, Timers, etc.).

3.2.3. *Tag down-chirp modulation*

3.2.3.1. *Operation principle*

One inconvenience of the backscattering technique is the mirrored harmonic generation on both sides of the carrier signal, which creates undesired interference across a wide range of frequencies. This issue has been addressed in the literature through a single-sided backscattering front-end that emulates a complex modulation signal $p(t)$ using four complex impedances (Iyer et al. 2016; Zhang et al. 2016c). In this section, a novel tag modulation based on LoRa backscattering is proposed, which achieves similar spectrum efficiency as single-sideband backscattering, but through a completely different approach. As will be seen in more detail, the proposed modulation in this section involves the unspreading of one sideband of the backscattered signal, thereby increasing its peak power. Simultaneously, the other sideband is spread by a factor of two, effectively disguising it as noise for the other channels. In practice, the double-spread sideband always remains below the noise floor.

The modulation described in this section stems from the demodulation process used in LoRa communications (Semtech Corporation 2015; Haritsa et al. 2020). LoRa encodes information circularly shifting a base chirp (up-chirp), which starts at f_0 and finishes at $f_0 + BW$. Equation [3.2] mathematically describes the base chirp signal $c(t)$ sent by the LoRa transmitter,

$$c(t) = e^{j2\pi(f_0 t + \frac{BW}{2T}t^2)}, \qquad 0 \leq t \leq T \qquad [3.2]$$

where BW is the bandwidth, T is the period and f_0 is the initial frequency of the chirp. To demodulate the transmitted chirps, the receiver multiplies the incoming signal by a complex conjugate time-reversed chirp $c^*(T - t)$ (down-chirp) with the same bandwidth and period as the transmitted chirp. The signal $r(t)$ resulting from multiplying the up-chirp and the down-chirp is described by [3.3]. We can realize that $r(t)$ perfectly matches with the equation of a complex sinusoidal signal $Ae^{j(\omega t+\phi)}$,

$$r(t) = e^{j2\pi((2f_0+BW)t+(2f_0-B)\frac{T}{2})}, \qquad 0 \leq t \leq T \qquad [3.3]$$

Since the load-switching process multiplies the incoming signal $c(t)$ by the switch modulation signal $p(t)$, the idea lies in directly modulating the switching frequency of the tag with a down-chirp signal with the same bandwidth and SF than $c(t)$, consequently bouncing the complex sinusoidal signal $r(t)$. The result of this operation is the backscattering of a frequency tone with the consequent processing gain. Figure 3.32 shows a diagram of the modulation process described in this subsection.

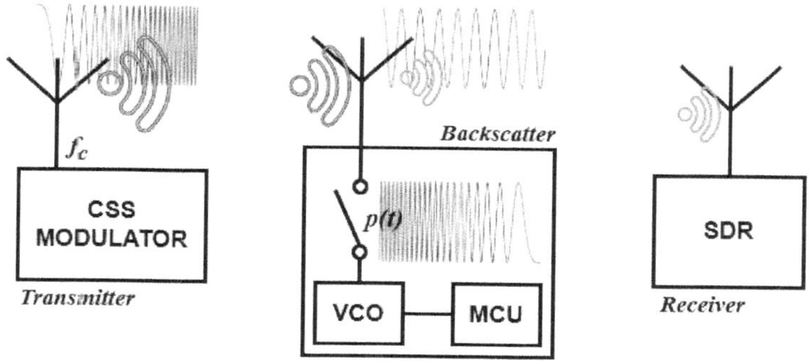

Figure 3.32. *Backscatter modulation diagram to bounce a sinusoidal tone from an incoming CSS signal*

In the conventional backscattering of CSS signals, this processing gain is obtained at the receiver side when the signal is demodulated, in contrast to the method proposed here, where the signal is directly backscattered from the tag already with the processing gain. The aforementioned processing gain can be determined mathematically by the ratio between the data rate of the spreading sequence R_c (chips/second), and the real data rate R_b (bits/second) of the communication, as described by [3.4],

$$G_p = 10 \cdot log_{10}(\frac{R_c}{R_b}) \qquad [3.4]$$

3.2.3.2. *Communication scheme*

At this point, the tag's information can be encoded in several ways. The simplest is to enable and disable $p(t)$ to generate an amplitude modulation. Alternatively, we can change the frequency of the backscattered tone by circularly shifting $p(t)$, thus creating an FSK modulation. Furthermore, by changing the slope of the down-chirp modulation – modifying BW or SF – to not match that of the carrier, another CSS modulation with a wider or thicker spectrum can be created. The range of possibilities is wide.

To encode the tag's information, the following method is proposed. Data is encoded implementing an OOK modulation over the backscattered frequency tone. Therefore, a digital one is transmitted when $p(t)$ is modulated with a down-chirp and the frequency tone is backscattered. On the contrary, cutting off the modulation $p(t)$, nothing is backscattered, which is equivalent to a digital zero. At this point, information could already be sent following the communication scheme presented in the previous section. However, to further increase the detectability of the tag, it is proposed to use a Barker sequence to encode the information and provide a second processing gain to the system. A Barker code is a binary sequence with the ideal auto-correlation property. This method consists of encoding individual bits with a digital sequence that the receiver unambiguously recognizes, which abruptly decreases the probability of error. This method is commonly used in direct sequence spread spectrum (DSSS) modulations. In the experiments performed in the laboratory, a 7-bit sequence -[1110010]- with a peak to side-lobe ratio of 16.9 dB is used.

Therefore, the proposed modulation achieves a double processing gain by first modulating the tag with a down-chirp, thereby reflecting a frequency tone with the first processing gain, and second, by encoding the tag's information with a Barker sequence, providing the second processing gain to the modulation. These processing gains depend on the resolution of the DAC and the length of the Barker sequence, respectively. Therefore, they can be adjusted depending on the application.

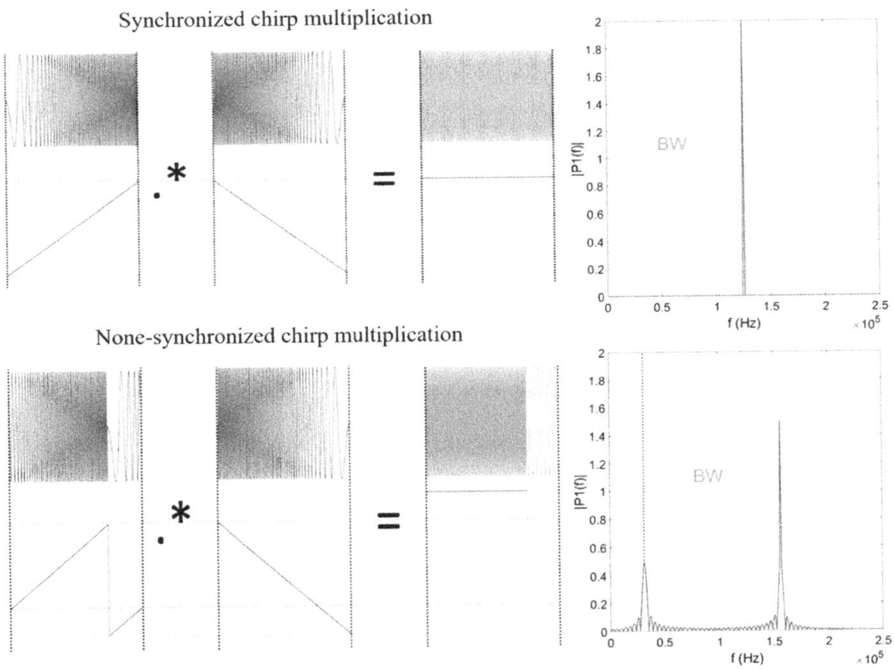

Figure 3.33. *Misalignment effect on chirp multiplication*

3.2.3.3. *Carrier-tag synchronization*

Since the backscattering tag does not integrate any receiver, there is no synchronization between the carrier LoRa transmission and the down-chirp generated in the tag. Figure 3.33 shows the effect of multiplying an up-chirp and a down-chirp miss-synchronized. Non-synchronized chirp multiplication generates two frequency tones with a separation between them equal to the chirp bandwidth (BW). The chirp bandwidth used in the simulation of Figure 3.33 has been set to 125 kHz. It can be observed that processing gain is distributed between the two frequency tones. Therefore, maximum processing gain is obtained when the up-chirp and down-chirp are synchronized, while the minimum is when one is shifted half of the chirp period relative to the other.

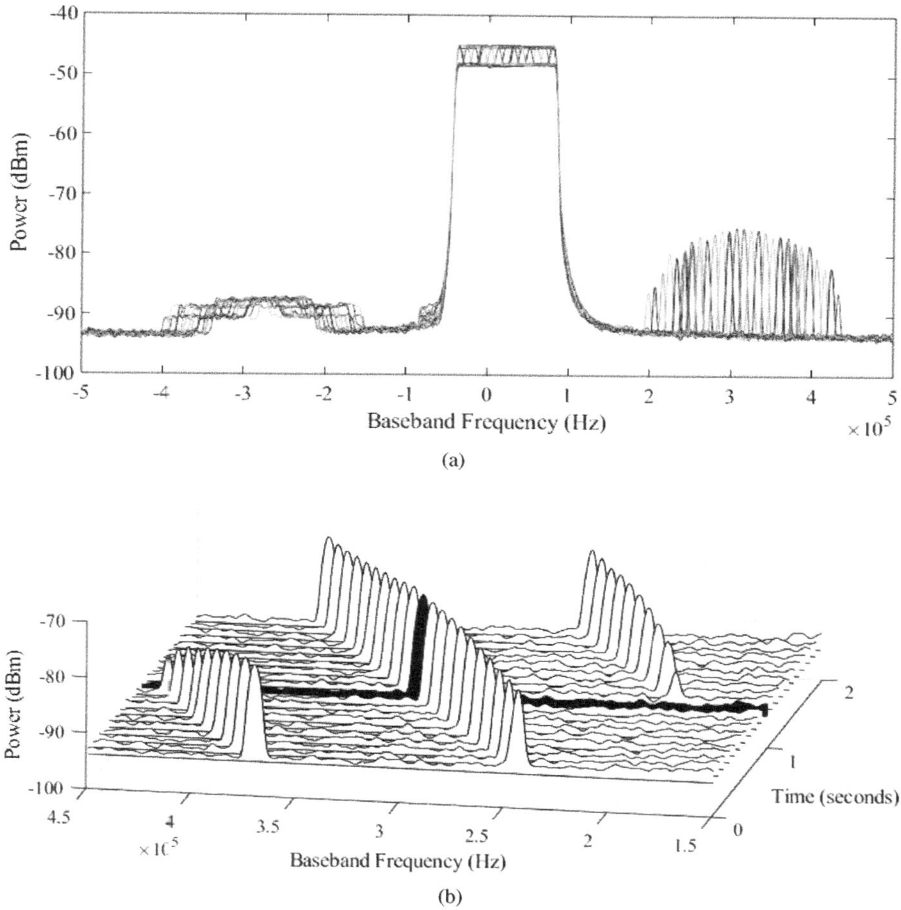

Figure 3.34. *Misalignment effect on chirp multiplication*

Figure 3.34a shows the measured spectrum when the down-chirp is deliberately shifted with respect to the impinging LoRa transmission. It clearly shows the processing gain distribution mentioned above. Figure 3.34b shows a 3D diagram of the measured right backscattered sideband, where the black-shadowed slide is the case when carrier up-chirp and tag down-chirp are perfectly synchronized.

This effect is difficult for the tag data demodulation, since the receiver does not know the frequency of the backscattered tone signal, which in fact can vary with each tag transmission when random waiting delays between packet transmissions are used. Beyond that, simple carrier or tag oscillator drift can shift the frequency of the backscattered tone. Since tag data is encoded using a known Barker sequence, a Matlab demodulation algorithm that performs the correlation between the measured IQ samples and the Barker sequence for all of the frequencies from $f_c+(f_p-(BW/2))$ to $f_c + (f_p + (BW/2))$ is implemented. The maximum correlation peak determines the frequency of the backscattered signal. By using lower-frequency increments (Δf) for the correlation sweep, the accuracy of peak identification can be enhanced at the cost of increased processing time. Once the backscatter frequency is determined, the complete packet data can be demodulated.

3.2.3.4. *Tag implementation*

The down-chirp has been synthesized in the tag with a voltage-controlled oscillator (VCO) controlled by the digital-to-analog converter (DAC) of the MCU. The tag prototype has been implemented using an 8-bit AtTiny417 MCU from Microchip, an LTC6990 VCO from Analog Devices and the already-mentioned ADG902. Figure 3.35 shows the prototype of the tag. It must be noted that the ability of the tag to synthesize the chirps depends on the resolution of the DAC. The total power consumption of the implemented tag is 635 μA and the cost is \$4.34.

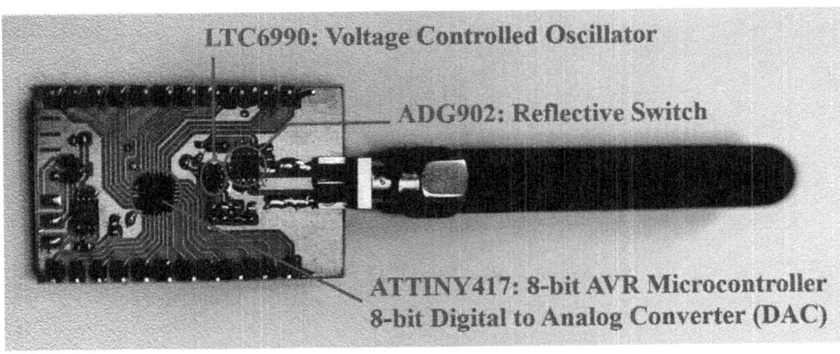

Figure 3.35. *Top view of the tag's prototype*

3.2.3.5. *Experimental results*

To compare the power peak of the reflected signal between the conventional OOK modulation and the down-chirp modulation proposed in this section, the resulting spectrum for both methods has been superimposed in Figure 3.36. For the experiments, the bandwidth (BW) and the SF of the LoRa transmitter were set to 125 kHz and 12, respectively. Finally, as observed in Figure 3.36, a processing gain of 13.41 dB was obtained.

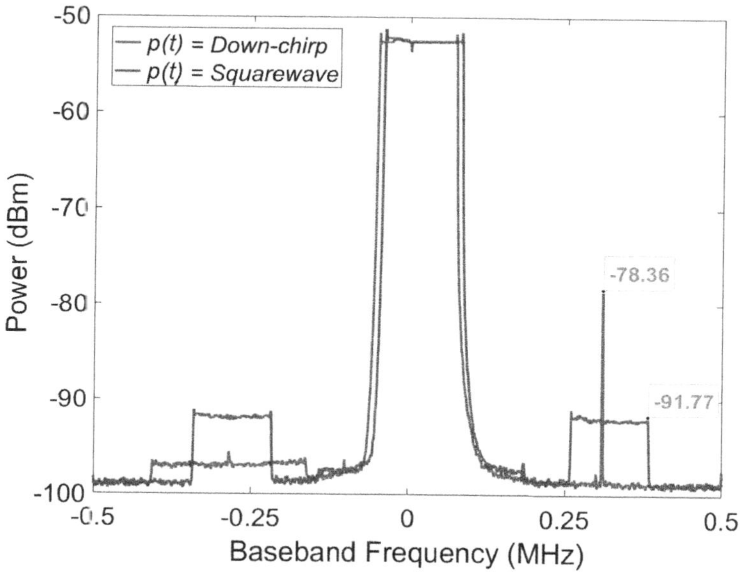

Figure 3.36. *Spectrum comparison between conventional backscatter and down-chirp modulated backscatter. Source signal: CSS modulation (up-chirp), $f_c = 868\ MHz$, $BW = 125\ kHz$ and $SF = 12$. Conventional backscatter signal (orange): squarewave, $f_p = 300\ kHz$, duty cycle = 50%. Down-chirp backscatter signal (blue): CSS modulation (down-chirp), $f_p = 300\ kHz$, $BW = 125\ kHz$ and $SF = 12$*

The measurements presented in this section were taken in bistatic mode. The receiver was placed at a distance of 20 meters from the transmitter, and the tag moved linearly along the distance between both. The power and the frequency of the transmitter were set to 20 dBm and 868 MHz, respectively. The tag's signal was recorded using the Adalm Pluto SDR and subsequently processed with Matlab. Figure 3.37 shows the IQ signal recorded with the SDR at a distance of 13 meters from the transmitter and 7 meters from the receiver.

Figure 3.38 shows the correlation of the incoming signal with the Barker sequence, where the demodulated data can be clearly appreciated. To test the communication, a fixed hexadecimal number -0x97- was transmitted by the tag.

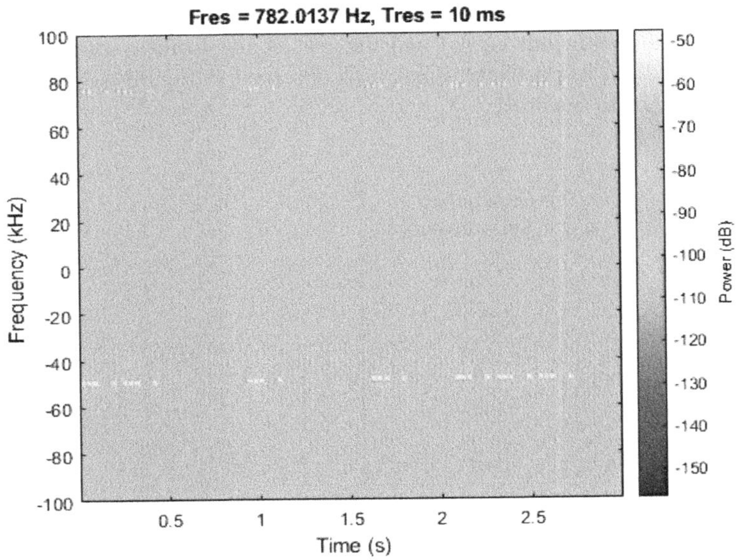

Figure 3.37. *IQ recording of the backscattered signal encoded with the 7-bit barker sequence*

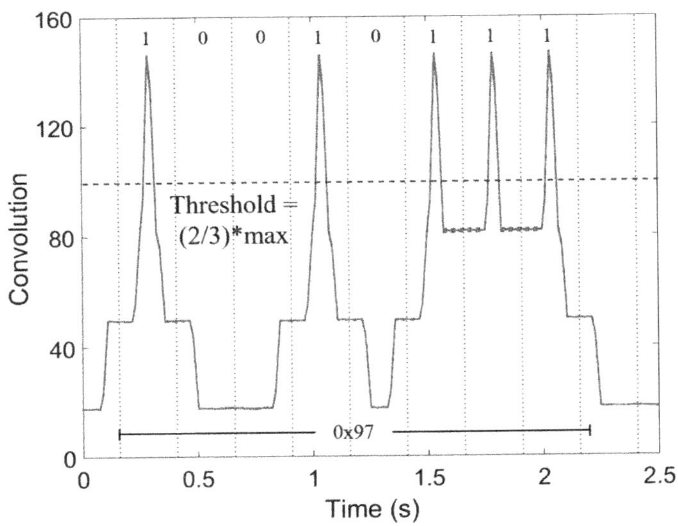

Figure 3.38. *Convolution of the backscattered signal with the Barker sequence. Tag demodulated data: 0x97, [10010111]*

The data rate of the communication, without considering the Barker code, can be calculated as a function of the chirp duration, which in turn depends on the BW and the SF. Each chirp encodes one bit. With the $BW = 125\ kHz$ and $SF = 12$ aforementioned, a data rate of $30.31\ bits/s$ can be achieved. Taking into consideration the Barker code, the data rate is reduced to $4.36\ bits/s$. However, these communication speeds are not the maximum rates. In fact, it is possible to modulate shorter chirps within the period of a base chirp, allowing more bits to be encoded in a single chirp. Higher modulation data rates and their effect over the range and consumption should be further explored.

Finally, it has been mentioned that the proposed down-chirp modulation presents two trade-offs in terms of processing gain: the down-chirp modulation gain can reach up to $25.33\ dB$ with a 12-bit resolution DAC, thereby increasing power consumption, while the Barker sequence gain can reach up to 22.3 dB of gain with a 13-bit Barker sequence, resulting in a reduced communication data rate.

3.3. Tag power consumption

This section addresses the energy consumption and provides a summary of the power drawn by each component used in the implementation of the aforementioned front-ends. The overall power consumption of the tag depends not only on the components used but also on factors such as operating frequency and biasing voltage. Table 3.3 summarizes the power consumption of each component used.

For the load modulation front-end, the microcontroller can be directly used to generate the switching frequency with the PWM functionality, or it can be used along with an oscillator only as a baseband digital modulator. The former configuration requires running the clock at $1\ MHz$ to generate the $300\ kHz$ switching frequency. On the contrary, the last configuration (MCU + LTC6907) allows running the MCU with the low-power $32\ kHz$ clock, only enabling or disabling the external oscillator as a function of the baseband data, making it a more efficient option. Despite using a prescaler to reduce the clock frequency from $1\ MHz$ to $\sim 300\ kHz$, the consumption still remains considerably higher than when using the $32\ kHz$ clock. Another consideration is that the 32 kHz clock works well in the LoRa packet-level OOK modulation due to its intrinsic low data rate, but is impractical for synthesizing down-chirps or higher data rate modulations.

From Table 3.3, it can be observed that the components have different minimum operating voltages. As mentioned in the amplified front-end section, when biasing voltages are not matched, the implementation becomes less efficient, since it requires using regulators, converters or, in the worst case, voltage dividers. For instance, let us consider the LTC6990 and LTC6907, which have minimum operating voltages of 2.25 *V* and 3 *V*, respectively. Assuming that the desired voltage is 1.8 *V* (the minimum of the microcontroller), one option is to study their behavior when biased below their minimum voltage. While it is not recommended, this technique would likely work with the LTC6990, since it is closer to 1.8, but may not be effective with the LTC6907, leaving only the power conversion option. Therefore, the power consumption of the regulator should be added to the previous table. Fortunately, regulators designed for voltage conversions between standard biasing voltages, such as in this case (3–1.8 V), typically exhibit high efficiency.

Component	**Frequency**	**Voltage**	**Current**	**Power consumption**
Backscattering front-end				
RF switch ADG902 (Analog Devices 2003)	868 MHz	1.8 V	<1 μA	~1 μW
RF switch ADG918 (Analog Devices 2016)	868 MHz	1.8 V	<1 μA	~1 μW
BJT HBFP405 (Infineon Technologies 2019)	868 MHz	1.8 V	280 μA	504 μW
Tunnel Diode AI301A (TekWiki 2025)	868 MHz	114 mV V	1.2 mA	136.8 μW
VCO LTC6990 (Analog Devices Linear Technologies)	237.5 - 362.5 kHz	2.25 V	180 μA	405 μW
Oscillator LTC6907 (Analog Devices Linear Technologies)	300 kHz	3 V	36 μA	108 μW
MAX9916 (Analog Devices Maxim Integrated)	-	1.8 V	36 μA	64.8 μW
MAX9634 (Analog Devices Maxim Integrated)	-	1.8 V	1 μA	~1 μW
DAC (AtTiny417) (Microchip Technology 2020)	-	1.8 V	96 μA	172.8 μW
Control unit				
MCU AtTiny402/417 (Microchip Technology 2021)	32 kHz	1.8 V	7 μA	12.6 μW
	1 MHz prescaled	1.8 V	455 μA	819 μW

Table 3.3. *Summary of power consumption for the components used in the implementations*

Nowadays, there are several energy harvesting techniques available to collect energy from sources such as light, electromagnetic fields, thermal gradients, movement and pressure, among others. Although RF energy harvesting is a commonly used technique in backscattering tags, it has not been studied in this book. This is due to two reasons: firstly, the minimum signal strength required to power up the tag limits the communication coverage, and secondly, the reflected power is considerably reduced, as demonstrated in section 2.2. Therefore, light energy harvesting has been considered a good option to power up the tag. The AEM10941 solar energy harvesting IC has been chosen due to its outstanding features. It integrates an ultra-low-power boost converter that can start harvesting energy from an input power as low as 3 μW at 380 mV. Furthermore, it offers a low-voltage output at either 1.2 V or 1.8 V, as well as a high-voltage output that can be configured between 1.8 V and 4.1 V. These outputs are driven by highly efficient, low-noise and high-stability low-drop regulators (LDO). This double output helps to cope with the difference in the minimum operating voltage of the front-end components.

Energy can be stored in various ways to meet application requirements, including Li-ion batteries, supercapacitors, or lithium-ion capacitors (LIC). Li-ion batteries are known for their high stored energy compared to supercapacitors, while supercapacitors excel in providing instantaneous power. LICs combine many of the advantages of both supercapacitors and Li-ion batteries. Figure 3.39 shows a 60 mAh LiPo battery, a 1.6 F LIC capacitor, a 1 F standard supercapacitor and a 220 mF CAP-XX prismatic supercapacitor. The CAP-XX brand offers the world's thinnest and highest power density supercapacitors, which are attractive due to their thin and lightweight package, as well as their very low ESR and leakage current. However, these cutting-edge supercapacitors come at a higher cost compared to standard supercapacitors and slightly higher than LIC supercapacitors. Ultimately, the choice of the energy storage solution depends on application requirements, considering factors such as cost and performance conditions.

(a) (b) (c) (d)

Figure 3.39. *Photography of several energy storage elements: (a) lithium polymer battery, (b) lithium-ion capacitor, (c) standard supercapacitor and (d) prismatic supercapacitor*

Figure 3.40 compares the power leakage over a period of 23 minutes for the aforementioned energy storage devices. It can be observed that, after 20 minutes, the LiPo battery, CAP-XX supercapacitor, LIC supercapacitor and standard supercapacitor discharged 0.10%, 0.67%, 3.25% and 9.83% of their respective capacities.

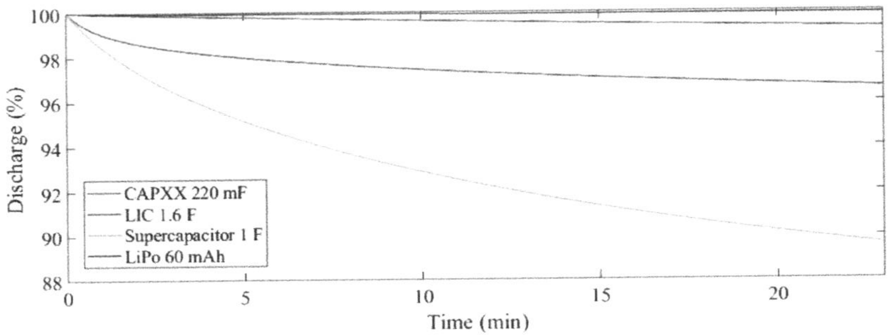

Figure 3.40. *Comparison of power leakage over time for a lithium polymer battery, a lithium-ion capacitor (LIC), a standard supercapacitor and a prismatic supercapacitor*

The light energy harvesting system has been tested using an indoor LL200-2.4-37 flexible solar cell from PowerFilm. This solar cell is capable of providing an open-circuit voltage (V_{oc}) of 2.2 V and a short-circuit current (I_{sc}) of 60 μA under 200 lux illuminance conditions, and V_{oc} of 2.9 V and I_{sc} of 200 μA under 1000 lux illuminance conditions. Indoor light levels typically range from 100 to 1000 lux (Mathews et al. 2015).

The harvested energy is stored in the mentioned 220 mF CAP-XX supercapacitor (part number: DMT3N4R2U224M3DTA0). This supercapacitor features an impressively low leakage current of only 5 μA and is packaged in a slim 2.2 mm form factor, making it an ideal choice for durable and low-profile tags. Figure 3.41 depicts the energy harvesting system, showing the flexible solar cell, the energy manager (AEM10941 from e-peas semiconductors) and the CAP-XX prismatic supercapacitor. This capacitor is capable of retaining energy throughout the night, ensuring continuous operation. By duty cycling the tag's operation, it is expected to function perpetually until component failure. One notable advantage of supercapacitors over lithium-ion batteries is their significantly longer lifespan. While lithium-ion batteries typically offer an average of 500–10,000 charge cycles, supercapacitors can range from 100,000 to a million cycles.

The setup depicted in Figure 3.41 has been used to power two different LoRa backscattering tags: one with the load-modulated front-end and another with the reflection amplifier front-end, with a power consumption of 582 μA and 984 μA, respectively. Figure 3.42 shows the measured voltage decrease of the CAP-XX 220 mF supercapacitor while powering the tags. The entire tag is powered at 1.8 V using the low-voltage regulated output. The LDO regulator is capable of continuously supplying the current required by the load-modulated tag for a duration of 10.75 minutes and the reflection amplifier tag for a duration of 5.67 minutes.

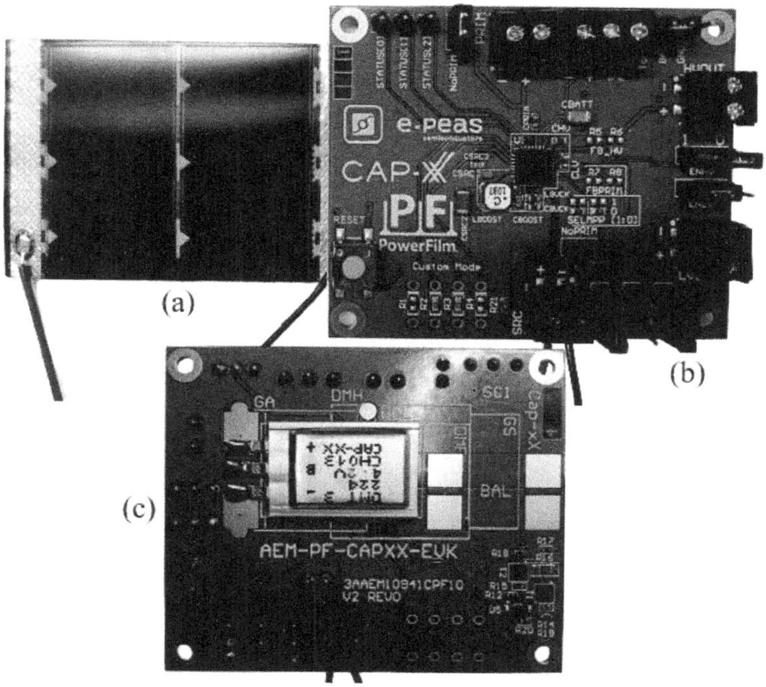

Figure 3.41. *Light energy harvesting system (AEM10941 Evaluation Kit): (a) flexible solar cell, (b) front view of the board and (c) rear view of the board with the CAP-XX supercapacitor*

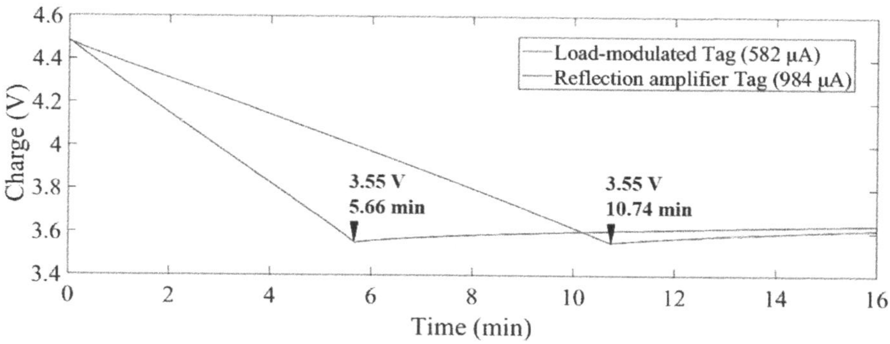

Figure 3.42. *Comparison of power leakage over time for a lithium polymer battery, a lithium-ion capacitor (LIC), a standard supercapacitor and a prismatic supercapacitor*

Considering the low-data rate modulation described in section 3.2.2, with a packet duration of 320 ms, the tag can transmit either 2015 packets or 1,063 packets (depending on the front-end used) without harvesting energy. Figure 3.42 reveals that when the supercapacitor voltage decreases to 3.55 V, the LDO regulator is no longer able to supply the power required by the tag, disabling the 1.8 V output. After disabling the tag power supply, it gradually starts harvesting light energy again.

3.4. Conclusion

From this chapter, several conclusions can be drawn about the tag's front-end design and the implementation of its modulation scheme, covering both the physical-layer modulation and the uplink communication protocol.

Different topologies have been studied for the implementation of the backscattering tag. On the one hand, the conventional load-modulated front-end has been examined. This system stands out for its extremely low power consumption, as it only requires an RF switch with minimal current consumption, making it suitable for passive implementations or applications with super-low power requirements. Subsequently, the integration of a reflection amplifier in the tag front-end has been studied for applications requiring longer communication ranges. The two main low-power reflection amplifier topologies found in the literature (tunnel diode-based and transistor-based reflection amplifiers) have been successfully designed, manufactured and characterized. The main advantage of the tunnel diode-based amplifier is its high gain and low power requirements, which has recently sparked significant interest. However, due to its low bias voltage, this high efficiency diminishes when integrated into a complete system, as logic circuits operate at much higher voltages, necessitating the use of converters to integrate them into the same tag. On the contrary, despite exhibiting modest efficiencies, the transistor-based reflection amplifier can operate at voltages that match those of microcontrollers, making the overall integration more efficient. Therefore, based on this premise, it can be concluded that the transistor-based reflection amplifier appears to be a better option for the design of backscattering tags.

Reflection amplifiers are pseudo-oscillators strategically designed to amplify without starting to oscillate and their behavior depends on the input power. Therefore, its integration into backscattering tags poses some challenges that have not been addressed in the literature. For input powers high enough, the reflection amplifier can start to oscillate, generating interferences in the spectrum and potentially disabling all neighbor communications, since the noise floor increases within the tag band. Furthermore, keeping the amplifier always turned on is not an efficient practice for energy management and reduces the tag's lifespan. Therefore, the integration of a current sensor to activate and deactivate the reflection amplifier as a function of the input power has been proposed, thus solving the two aforementioned issues.

For the communication implementation, two different backscattering modulations have been presented, along with an uplink communication scheme. Each modulation exhibits its own advantages and weaknesses. The OOK LoRa-packet modulation stands out for its low power consumption and compatibility with commercial off-the-shelf LoRa receivers for demodulation. However, due to the backscatter information being piggybacked by LoRa packets, the achieved data rate is not particularly high.

On the other hand, the down-chirp modulation implements a more optimal spectrum utilization and takes advantage of the processing gain inherent to LoRa modulation. Furthermore, an additional processing gain is achieved by encoding the tag information with a 7-bit Barker sequence. However, this modulation requires more power due to the integration of a VCO and the higher MCU clock speed required for synthesizing the LoRa down-chirps, and it must be demodulated using an SDR. Since the tag information is demodulated with an SDR, further work could be done exploring the maximum communication data rate as a function of the tag power consumption.

Finally, the tag power consumption has been summarized based on the elements used in the hardware design. Furthermore, a light energy harvesting system has been assessed to power up the tag, using an indoor solar cell, and a low-profile, low-leakage current, prismatic supercapacitor as energy storage element.

4

Backscattering for Localization Applications

This chapter examines the fundamental concepts required to implement a localization system based on the LoRa signal backscattering technology detailed in this book. It bridges the theoretical foundations presented in earlier chapters with practical implementation, demonstrating just one of this technology's countless potential application domains.

4.1. Introduction

In recent years, indoor localization systems have generated great interest due to their potential applications (Xiao et al. 2016; Yassin et al. 2016; Zafari et al. 2019). Most of the methods proposed in the literature, as well as commercial systems, focus on improving localization precision (Seco et al. 2009). In some applications, such as autonomous robot guidance (Raghavan et al. 2010), the precision of localization is fundamental because it requires an exact location in a coordinate system. Performance depends greatly on the method used: time of arrival (ToA), time difference of arrival (TDoA), angle of arrival (AoA) or received signal strength indicator (RSSI), and it also depends on the environment, attenuation introduced by objects, movement and the localization algorithm used. Depending on the localization method, different wireless systems have been explored in the literature: Wi-Fi, Bluetooth, ultra-wideband (UWB), modulated continuous-wave radar (FMCW), RFID, etc. LoRa-based localization systems have recently been proposed for indoor and outdoor environments (Lam et al. 2018, 2019; Li et al. 2018; Lin et al. 2019). Some of these systems were compared in Landt (2005) and Finkenzeller (2010). Their precision often increases with the number of anchors (Moravek et al. 2011, 2013) and these systems require LoS (Line-of-Sight) scenarios (Čabarkapa et al. 2015).

Another interesting group of localization techniques includes those that are based on fingerprint classification. In this case, the system estimates the location of an object by matching the RSSI measurements with the closest location fingerprints. RSSI measurements are often obtained from Wi-Fi access points (Yang et al. 2012) or Bluetooth beacons (Tegou et al. 2018; Li et al. 2019) in indoor applications or Sigfox, LoRaWAN (Ha et al. 2019) or mobile communications in large areas (Aernouts et al. 2018). However, these methods require a calibration of the system by recording the exact location on a map. In these systems, two main groups of classification techniques have been reported: neural networks (Gogolak et al. 2011; Basiouny et al. 2017) and machine learning techniques (Del Mundo et al. 2011). The main drawback of previous localization methods is that the large number of anchors and the manual calibration procedure decrease the system's return on investment (ROI). However, in certain applications, localizing the target within an approximate area, such as a room, is sufficient (Martin et al. 2014; Jia et al. 2015; Tegou et al. 2018). Within the various medical applications, monitoring activities of daily living (ADL) (Tiusanen 2013; Blackman et al. 2016; Debes et al. 2016; Susnea et al. 2019) constitute an example of the second group. A major challenge for these applications is that the tracking devices must be noninvasive and private because they are intended to be used by the elderly. Therefore, devices must be worn, and the battery lifetime must be long enough to avoid continuous recharges that can cause the user to stop using the system. Both the installation of the localization system and the cost mean that a reduced number of anchors, capable of covering large areas (e.g. care facilities or hospitals), have to be installed. From a practical perspective, it should be noted that in most indoor localization methods, when the number of anchors used is small (or when the cost of installing several anchors is prohibitive), the precision is approximately several meters (Li et al. 2019). In other words, they can be used to roughly recognize a room.

The main objective of this chapter is to study the performance and viability of an indoor room-level localization system based on LoRa backscattering. This application aims to contribute to the improvement of human and social welfare through the application of the system in fields such as patient management in hospitals and care facilities, capacity monitoring in establishments of different types, autonomous buildings, etc. Compared to other systems based on active beacons (e.g. Bluetooth, Wi-Fi or UWB), LoRa backscatter stands out for its lower power consumption and long range as a result of the high sensitivity of LoRa receivers. This LoRa backscatter device can be embedded in a wristband or any other wearable device, thereby obviating the need for heavy and uncomfortable electronics. The localization system is made up of several LoRa receivers distributed in different rooms, a LoRa transmitter located at a central point and the backscatter device, which is carried or worn by an individual. Commercial low-cost LoRa modules have been used in the experiments to validate the system. A simple algorithm based on the reception of the highest signal strength is used to determine the position (the room where the user is). In order to improve accuracy, different machine learning algorithms are compared. In these cases, algorithm training is accomplished by acquiring a set of samples while the user moves randomly within each of the rooms. The acquisition of random points greatly accelerates the training procedure. It is only necessary to know which room the samples have been taken in, and not the coordinates of the receivers or the transmitter. Parameters such as transmitter power and propagation model are also not necessary to program the algorithms. This simplifies the training procedure and practical implementation in real environments.

The chapter is organized as follows. Section 4.2 describes the system, including the proof-of-concept prototype for the LoRa backscatter. A propagation model is proposed to verify the algorithm based on the simulated data. The simulated results are described in section 4.3. Experimental results are presented in section 4.4. Section 4.5 compares the features of the proposed room-level localization technique with other localization technologies. Finally, section 4.6 summarizes the conclusions.

4.2. System architecture overview

The basis of the backscattering device used for this application is described in Chapters 2 (theory) and 3 (implementation). The tag backscatters LoRa packets to an adjacent channel using the load-modulation front-end and the OOK coding scheme described in section 3.2.2. The LoRa backscatter stands out for its low power consumption compared to other beacons because it does not require generating RF signals or processing the incoming LoRa packets.

A set of LoRa receivers, tuned at the backscatter adjacent channel, receive the data packets and measure the RSSI and signal-to-noise ratio (SNR) for each one. Each room has its own LoRa receiver, so the location of the tag can be identified by comparing the RSSI of the backscattered packet of the receivers located in each room. The receiver with the highest RSSI level determines the location of the backscatter device. Advanced fingerprint algorithms can be used to improve the decision. A diagram of a possible scenario is shown in Figure 4.1.

For system implementation, several low-cost LoRa32 development boards, TTGO brand, are used. These boards integrate an Espressif ESP32 microcontroller and a Semtech SX1276 transceiver. The ESP32 provides Wi-Fi and Bluetooth connectivity, while the SX1276 enables LoRa connectivity, making it a versatile triple wireless technology transceiver. The SX1276 transceiver can operate in a frequency range of 137–1020 MHz, covering the European, North American and Asian ISM bands. An advantage of the SX1276 transceiver is its high tolerance to frequency deviations, which allows packet demodulation with significant frequency shift. Moreover, in the rare case that the data cannot be demodulated due to considerable frequency shift, the SX1276 includes automatic frequency correction (AFC) functionality to adjust the reception frequency. The system implementation has been performed with the

European/American version of the LoRa32 board, which operates in the 868 MHz ISM band. The SX1276 allows us to set up several parameters of the radio and the LoRa packets. Experiments have been carried out with the default transmission power (17 dBm). However, if necessary, it can be adjusted from 2 dBm to 20 dBm. The cost of LoRa32 boards is less than 10 $, and no hardware modification is required, making the full deployment of the system cheap and simple.

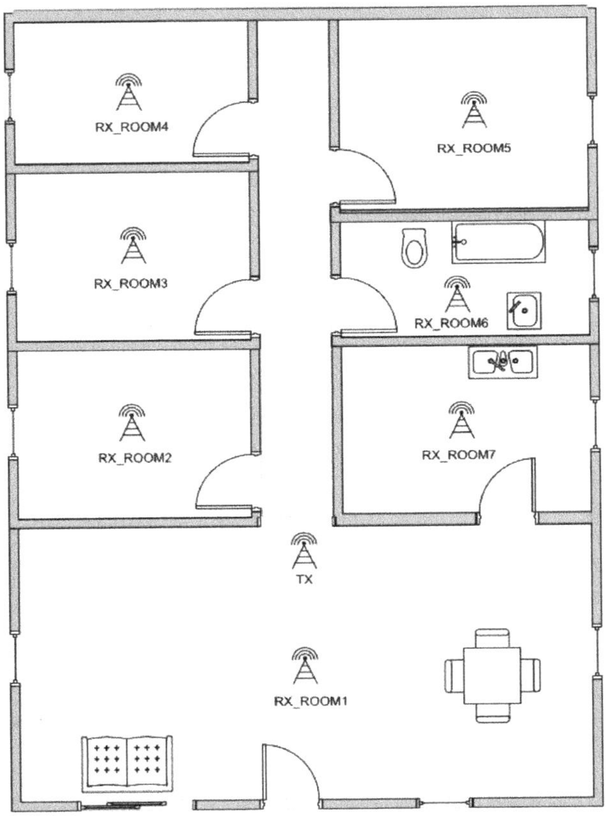

Figure 4.1. *Example of typical Tx and Rx placement in a house*

A Wi-Fi link is used to send the data to a cloud database. Alternatively, the data can be transmitted among nodes via point-to-point LoRa links or directly to a LoRa gateway, which uploads the data to a cloud database. This second method can be used in industrial environments or in buildings without coverage in some areas or without any coverage at all. Data are transferred to the cloud via the MQTT protocol (Message Queue Telemetry Transport). Each receiver publishes the RSSI and SNR measurements in an MQTT topic. To avoid errors arising from lost packets or uploading delays, the transceiver sends each packet with an identifier, which might be just a counter. This makes it possible to synchronize the data collected from the receivers. After the tag transmission, each receiver publishes a message to the MQTT broker with the packet identifier, the room identifier and the measured RSSI and SNR.

4.2.1. *Backscatter prototype*

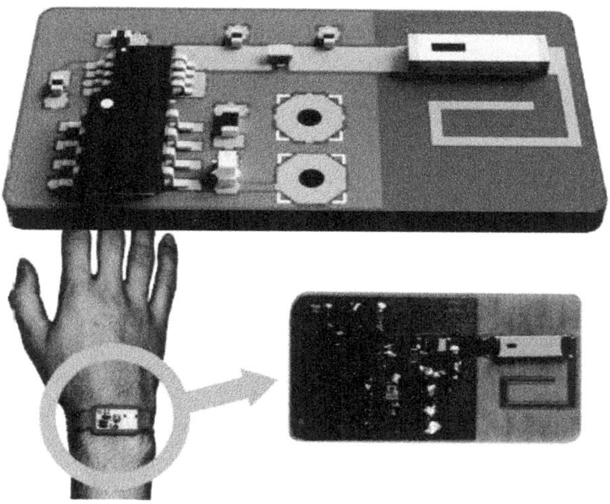

Figure 4.2. *Block diagram of the backscattering system (top) and photograph of a prototype (bottom). For a color version of this figure, see www.iste.co.uk/lazaromarti/low-power.zip*

For the tag prototype, the LoRa packet-level OOK modulation described in Chapter 3 is used. The implementation diagram is the one shown previously in Figure 3.34a of Chapter 3. This modulation has been chosen for two reasons: to keep the tag's power consumption as low as possible and to enable demodulation with commercial LoRa receivers, thereby reducing the overall implementation cost. Figure 4.2 shows a rendering of the PCB and an image of the manufactured prototype, which has been mounted on a 3D printed wristband. The tag integrates a low-profile wideband ceramic antenna (P/N: 0900AT43A0070) from Johanson Technology, designed to operate in the 868 and 915 MHz bands. The antenna consists of a ceramic monopole and an additional printed line used to adjust the frequency of the antenna in the 868 MHz band. To enhance impedance matching, an LC matching network with discrete components has been designed with the aid of the VNA. To switch between the open-circuit and the short-circuit states, the already mentioned ADG902 is used, and the control signal is generated directly using the PWM function of the AtTiny402 microcontroller. The microcontroller generates the tag identifier baseband data transmitted via backscattering. The total current consumption when the microcontroller wakes up, working at a 1 MHz clock speed with an operating voltage of 3.3 V, is 890 μA. It can be reduced even more by reducing the operation voltage, prescaling the clock frequency or integrating an oscillator. However, for the proof-of-concept, it has been implemented with the above-mentioned configuration. This stated power consumption is considerably lower than that of Bluetooth LE modules that typically require 7–15 mA when they are in reception or transmission states. An interesting aspect to mention is that the backscatter device has no transmission or reception mode, since it is not capable of processing or generating RF signals. Therefore, the tag periodically backscatters the identifier by reflecting the LoRa packets, and it is the receiver's task to obtain this information.

4.2.2. *Propagation model*

The method proposed consists of comparing the received signal strength of the backscattered packets among the receivers in each room, which are tuned in the sideband channel, where the tag transmits. In this section, an empirical model for performing simulations of coverage area and localization rates is proposed. The power received (P_R) can be expressed as in an RFID system using the radar equation (Nikitin et al. 2007):

$$P_r = \frac{P_t G_t}{4\pi d_t^2} \cdot \Delta\sigma \cdot \frac{1}{4\pi d_r^2} \cdot \frac{\lambda^2}{4\pi} \cdot G_r \qquad [4.1]$$

where P_t is the transmitted power, G_t is the gain of the transmitter antenna, G_r is the gain of the receiver antenna, $\Delta\sigma$ is the differential radar cross-section of the tag, and d_t and d_r are the distance from the transmitter to the tag and from the tag to the receiver, respectively. The receiver power increases when the tag is close to the transmitter or to the receiver. The locus of points with constant received power approximately describes an ellipse. However, the Friis model [4.1] is only valid in free space or in situations where the antennas have high directivity, which avoids multipath interference. In order to include multipath propagation, an empirical model is often considered. The RSSI is the power received expressed in dBm (Lazaro et al. 2020b):

$$P_r(dBm)(d_t, d_r) = P_r(d_0, d_0) - 10n_1 log(\frac{d_t}{d_0}) - 10n_2 log(\frac{d_r}{d_0}) - L_{wall} + X \qquad [4.2]$$

Equation [4.2] is split into two parts considering the two existing propagation paths: transmitting to the tag and backscattering to the receiver. In equation [4.2], d_0 is the reference distance (i.e. the midpoint between the transmitter and receiver $d_0 = (d_{Tx-Rx}/2)$), $P_r(d_0, d_0)$ is the average received power in d_0, and n_1 and n_2 are the path loss exponents, whose values depend on the environment and the height of the antennas. Typical values for indoor environments vary between 2.5 and 3. L_{wall} are the losses due to diffraction and attenuation of the walls, and X is a random variable that takes into account attenuation due to multipath propagation. $P_r(d_0, d_0)$ can be determined experimentally through the regression analysis of the measured data and depends on the transmitted power and the gain of the antennas. It is also a function of the height of the antennas and depends on the diffraction. However, in this section, the expression [4.2] derived from the Friis transmission equation is considered to perform the simulations.

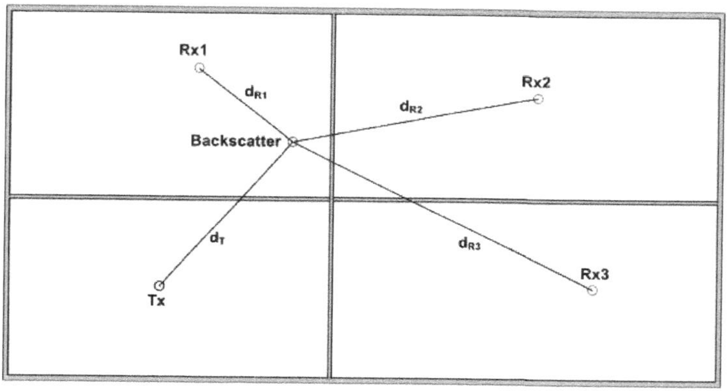

Figure 4.3. *Diagram used for the description of the propagation analysis*

A typical deployment scenario is shown in Figure 4.3. The attenuation due to the path loss between the transmitter and the tag is the same in all receivers, but the power of the received signals depends on the distance between each receiver and the backscatter $d_{r(i)}$ and the number of walls that each signal goes through. It is expected that the maximum received power will be that of the receiver that is in the same room as the backscattering tag because it will be closer to it and attenuation due to the walls will not be an issue.

Typical wall attenuation at 900 MHz for a brick wall is 5 dB and 15-18 dB for a double reinforced concrete wall (Pena et al. 2003). In the model, the distribution presented by the random variable X can be considered of log-normal type $X(dB) \sim N(0, \sigma)$, where σ (dB) is the standard deviation and the experimental results yield values between 2 and 4 dB for a single path (Lazaro et al. 2009) (in this case, a double path due to the backscatter channel is expected). The coverage probability (Prob) is that in which the average received power or RSSI (P_r) is higher than the receiver sensitivity S (dBm). For a log-normal distributed channel, this is given by Goldsmith (2005):

$$Prob(P_r > S) = \frac{1}{2} erfc(\frac{S - P_r}{\sqrt{2}\sigma}) \quad [4.3]$$

where $erfc$ is the complementary error function. For a σ equal to 8 dB and a received power of 10 dB above the sensitivity (fading margin), the coverage probability is about 90%.

The sensitivity of the LoRa receiver also depends on the noise figure of the receiver and the bandwidth (BW). It is calculated from the noise floor plus the required SNR (Semtech Corporation 2020):

$$S(dBm) = -174 + 10log(BW) + NF(dB) + SNR(dB) \qquad [4.4]$$

where BW is the bandwidth of the LoRa signal and NF is the LoRa receiver noise figure. The SNR is negative and can be approximated as a function of the spreading factor (SF), which can be selected on LoRa transceivers from 7 to 12 (Semtech Corporation 2020):

$$SNR(dB) = -7.5 - 2.5(SF - 7) \qquad [4.5]$$

The already mentioned high sensitivity of LoRa transceivers makes this wireless technology perfect for increasing the probability of backscatter tag detection in indoor environments, thus compensating for the typical attenuation experienced in such spaces, due to walls and multipath interferences. From equation [4.4], for an SF of 12, a bandwidth of 125 kHz and assuming a typical noise figure of 6 dB, the receiver sensitivity is -137 dBm.

4.2.3. *Room-level classification*

The classification process is schematically described in Figure 4.4. The LoRa transmitter regularly transmits packets using a counter, which is required for packet identification. The transmission time interval (packets transmitted per second) can be configured according to the desired tracking frequency or application. When it strikes the device, the signal coming from the LoRa transmitter is backscattered towards the receivers located in each room to identify the signal. Each receiver is tuned to the shifted frequency channel. The RSSI and SNR measured by the LoRa receivers and the packet identification number are sent to the MQTT broker. In this case, a Wi-Fi transmission is used because the ESP32 has a built-in Wi-Fi transceiver. However, in the absence of Wi-Fi coverage or capabilities, the receiving node can be configured to transfer the message via LoRa to a gateway with an internet connection, which will upload the data to the MQTT broker. In the experiment, a custom Mosquitto broker is used, but other MQTT brokers can also be used (Light 2017). There are two modes of operation: the training mode and the normal operation mode. In the former, the data are stored in an SQLLite database for later use in the training of the classifiers. Then, to train the system, data are provided together with the room where the backscatter is, allowing the system to learn the footprint of each backscatter location. To indicate the room where the backscatter is located during the training stage, a message is sent to an MQTT configuration topic using a mobile app. After that, the training data collected are used to train the machine learning model. The MQTT client is implemented in Python using the Eclipse MQTT Paho module. The scikit-learn module is used for machine learning (Kramer 2016). In normal operation mode, the room number is determined by the measured RSSI and SNR data, which are obtained from the subscription to the respective MQTT topics of the receivers. If any of the receivers do not receive the backscattered packet because they are outside its coverage area, the minimum value of RSSI (sensitivity) and SNR will be assigned. The simplest classification method consists of getting the room index of the receiver with the highest SNR or RSSI. As the resolution in the RSSI measurement is only 1 dBm for the LoRa transceiver used in this work, the use of the SNR measurements seems to be the most appropriate option due to their higher resolution, which tends to yield better results. In any case, the room can be determined with both the RSSI and the SNR measurements, providing the system with a double-check calculation. If advanced machine learning algorithms are used for room classification, the room index is obtained by running the trained classifier. The simulated and experimental results with different classifiers are described in the next sections.

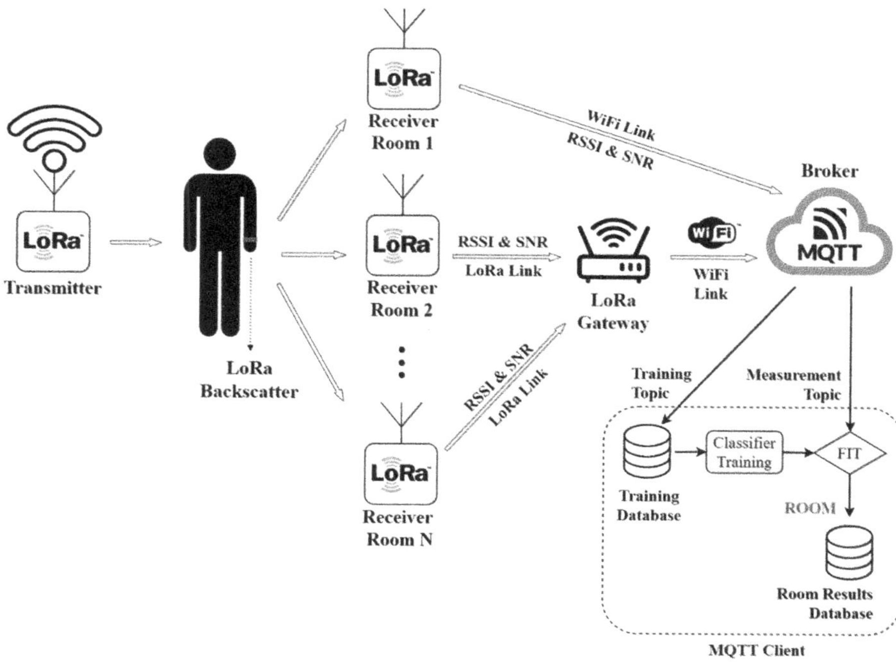

Figure 4.4. *Block diagram of the classification procedure. For a color version of this figure, see www.iste.co.uk/lazaromarti/low-power.zip*

4.3. Simulations

Figure 4.5 shows the power received in dBm calculated using the models [4.1] and [4.2] as a function of the location of the tag for a frequency of 868 MHz. A distance of 50 meters between the transmitter and receiver is considered. Figure 4.5a shows the ideal free-space case, whereas Figure 4.5b shows a scenario with decay factors (n_1 and n_2) of 2.5. The EIRP of the transmitter is 20 dBm, and the gain of the receiver antenna is 2.2 dBi. The gain of the tag antenna is -1 dBi (the typical surface-mounted antenna gain). In the simulations, the measured tag gain and the ideal square-wave waveform are used. An indoor scenario resulted in a considerable reduction in receiver power. However, these simulation results show that a LoRa backscatter can be detected within an area of about 100 m^2, although the attenuation of walls or other objects would decrease this coverage area. The experimental results measured in a house are presented in the next section.

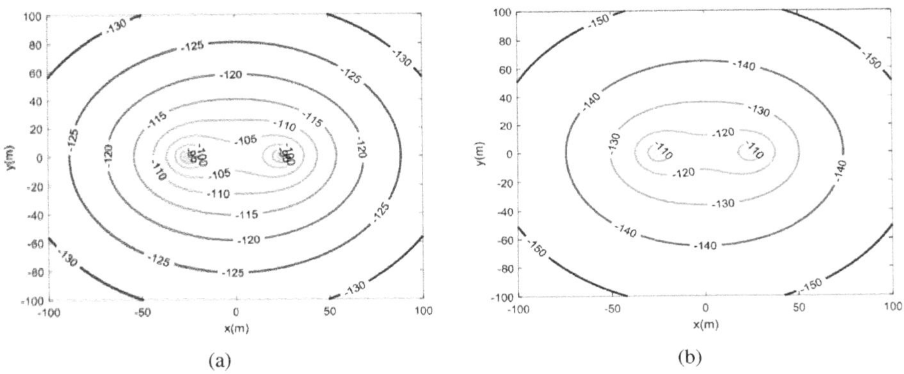

Figure 4.5. *Contour map of the received power as a function of the backscatter location (a) in free space and (b) with multipath considering a propagation decay factor of 2.5. For a color version of this figure, see www.iste.co.uk/lazaromarti/low-power.zip*

A series of simulations have been performed to validate the proposed method. Six rooms of different sizes, covering an area of 40 m × 40 m, are considered. The power received from model [4.2] is shown in Figure 4.6. A value of 10 dB is taken for wall attenuation with a standard deviation of 8 dB. The transmitter is located approximately in the center of room one.

Several classifiers can be used to determine the room number associated with the RSSI measured for each receiver. The simplest method consists of determining the room number through the receiver with the best power reception. Figure 4.7 shows the classified rooms for the last simulated scenario using the simplest RSSI classifier, where the room is assigned to the receiver that collects the strongest RSSI signal. This method has the advantage of not requiring a training step. Although this method worked reasonably well in the presence of noise, other classifiers have also been analyzed. Classifiers implemented in the Python scikit-learn machine learning module (Pedregosa et al. 2011) have been used for comparison. The classifiers are trained with 20% of the samples (simulated reception powers for several locations of the tag). The samples have been chosen randomly, and the other 80% is used to test the trained classifier.

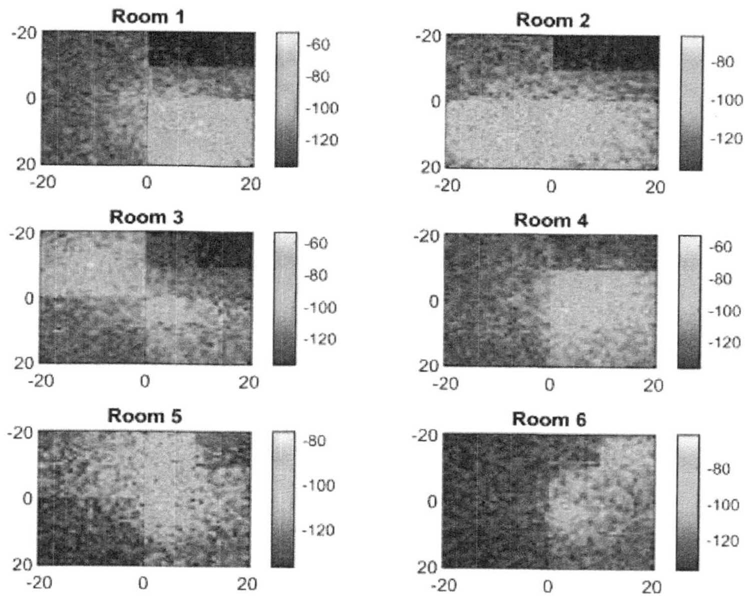

Figure 4.6. *Contour map of the received power as a function of the backscatter location for each receiver location. For a color version of this figure, see www.iste.co.uk/lazaromarti/low-power.zip*

Table 4.1 compares the accuracy obtained for each classifier. The best classifier found is the linear discriminant analysis (LDA), which achieved an accuracy of up to 91.5%. Table 4.2 shows the confusion matrix C for the LDA classifier. Each element of this matrix C_{ij} is equal to the number of observations known to be in room i and predicted to be in room j. For example, in the first row, the first column indicates the room where the backscatter is, in this case, room one. The algorithm has succeeded in its predictions in 93.84% of the cases, but in 3.07% of the cases, it has determined that the backscatter was in room two; in 1.54%, it has predicted room three, and in 1.54% room four. The results shown in Table 4.2 have been normalized by the total number of samples. Note that to improve readability, the values equal to zero have been left blank. As the table shows, there is a small degree of confusion with neighboring rooms.

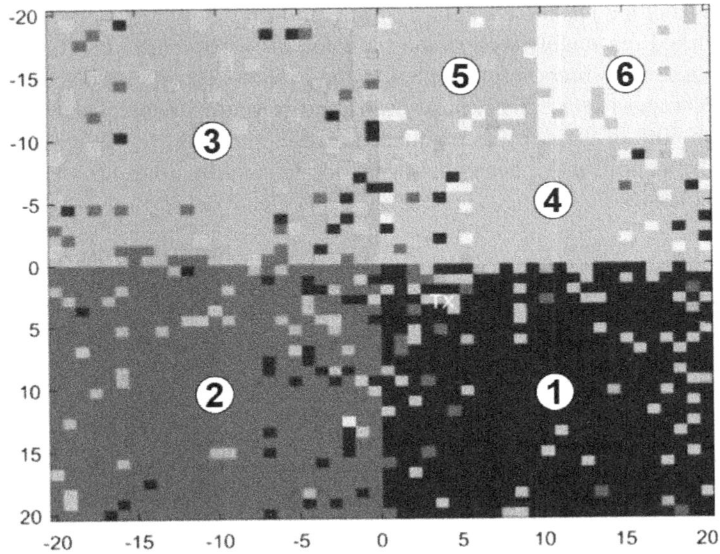

Figure 4.7. *Color map that shows the classified room using an RSSI level classifier. Each room is represented by a different color. The position of the receiver is indicated with the room number. The position of the transmitter is also shown. For a color version of this figure, see www.iste.co.uk/lazaromarti/low-power.zip*

Room	HRL	LR	LDA	KNN	CART	NB
1	84	93	94	90	88	90
2	87	91	91	94	89	90
3	85	96	97	93	91	93
4	80	80	90	80	77	82
5	85	72	83	90	79	83
6	92	54	94	89	97	94
Average	85.6	89.3	92.5	91.5	87.0	89.3

Table 4.1. *Accuracy (%) as a function of classifier*

Room	1	2	3	4	5	6
1	93.84	3.07	1.54	1.54	0	0
2	3.91	91.40	4.69	0	0	0
3	0	0.85	96.58	2.56	0	0
4	4.92	0	3.27	90.16	0	1.64
5	0	0	6.89	3.45	82.75	6.89
6	0	0	0	0	5.71	94.28

Table 4.2. *Normalized confusion matrix (%) using linear discriminant analysis*

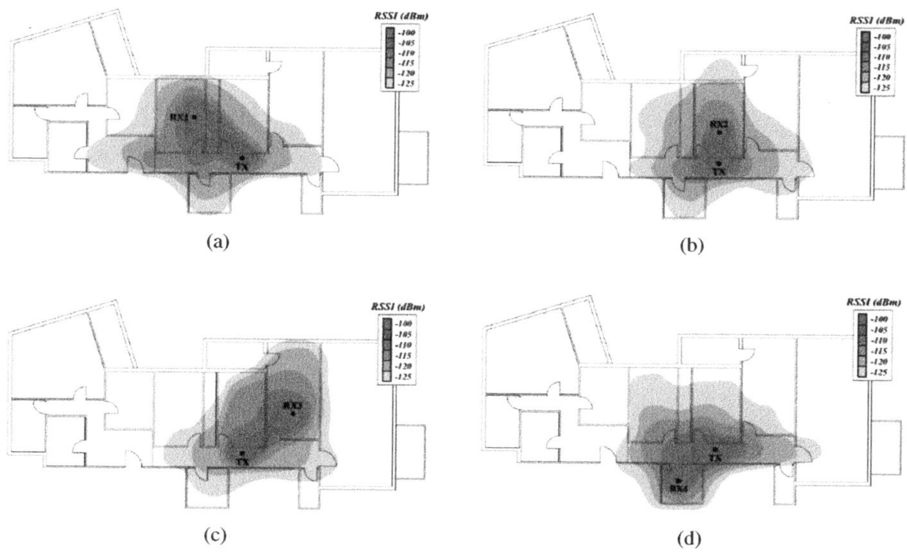

Figure 4.8. *Measured heatmap of the received power as a function of the backscatter location for (a) receiver RX1 located in room number one, (b) receiver RX2 located in room number two, (c) receiver RX3 located in room number three and (d) receiver RX4 located in room number four (d). For a color version of this figure, see www.iste.co.uk/lazaromarti/low-power.zip*

4.4. Experimental results

The proposed room-level localization system is tested in a complex scenario such as a house. Several experiments have been conducted to determine the coverage of the LoRa backscattering system. Figure 4.8 shows a heatmap of the RSSI received for the different rooms equipped with receivers. The backscatter is placed on the wrist of a volunteer who moves along different points on a predefined grid. The received signal strength and SNR at each point are then used to plot the heat map. The transmitter is located approximately in the center of the house (in the hallway) and a receiver is located in another room separated by walls. The receiving power strength on the receiver increases when the backscatter is near the transmitter or receiver. Reasonable coverage (with an RSSI greater than -122 dBm) is achieved in all situations. Therefore, the backscatter signal can be received by the nearest receivers. In this test, the transmitted channel frequency used is 868 MHz and the receivers are tuned to the right sideband channel shifted 300 kHz from the carrier. The LoRa transmission bandwidth has been set at 125 kHz and the SF is set to 12.

Room	HSL	LR	LDA	KNN	CART	SVM
1	82	82	93	90	82	88
2	89	86	83	81	81	88
3	40	97	97	93	93	93
4	56	86	86	82	82	84
Average	66.7	87.7	89.7	87.7	84.5	88.2

Table 4.3. *Accuracy (%) as a function of classifier*

Room	1	2	3	4
1	92.5	5.0	0.0	2.5
2	7.1	83.3	4.7	4.7
3	0.0	1.7	96.6	1.7
4	10.0	4.0	0.0	86.0

Table 4.4. *Normalized confusion matrix (%) using linear discriminant analysis (LDA)*

The system has been applied to the scenario shown in Figure 4.8. The transmitter (Tx) is located in the center of the hall, and four receivers (Rx_1,Rx_2,Rx_3,Rx_4) are installed in the adjacent rooms. The training procedure is performed as described in section 4.2. A comparison of the accuracy achieved in each room is summarized in Table 4.3. Considering the method based on the highest SNR level, the degree of accuracy is very good for rooms one and two because the received strength levels are high, whereas in rooms three and four, except for some points, the degree of coverage is much lower, or even nonexistent. Applying the same analysis, the accuracy for rooms three and four is worse compared to that for rooms one and two. This explains the low degree of accuracy for these rooms using the simplest method. However, the other classifiers are able to achieve high degrees of accuracy in all four rooms, regardless of the room's potential coverage. Experimental results show that the best average accuracy is obtained with the LDA classifier, although the accuracy for room two is worse than that obtained with the other classifiers. In any case, the differences in average accuracy among classifiers are slight, excluding the highest SNR-level classifier. The Gaussian Naïve Bayes (NB) classifier has not been analyzed because the measured data in the rooms did not meet the Gaussianity conditions for the method that would allow it to be applied. The normalized confusion matrix for the KNN classifier is shown as an example in Table 4.4. The elements of the principal diagonal that present the highest percentages correspond, as expected, to the neighboring rooms. There are a significant number of cases (10%) that increase confusion in the classification process between rooms four and one. This is because the coverage area of receiver 1 and receiver 4 partially overlapped with similar received levels.

4.5. Discussion

This section examines existing indoor localization technologies described in the literature. Although several surveys (e.g. Zafari et al. (2019) and Nessa et al. (2020)) on localization are available, this section aims to compare the advantages and limitations of the system proposed here in relation to other similar systems. Table 4.5 summarizes some of the properties, advantages and disadvantages of the technologies studied. The different technologies analyzed are described below:

– *Wi-Fi-based localization:* Wi-Fi networks (based on IEEE 802.11 standards) operate on ISM bands and are used to provide internet access to computers and smartphones, especially in indoor environments. Therefore, Wi-Fi is an ideal system for indoor localization and has been widely examined in the literature. Existing access points deployed for communication can be used as reference (anchor) nodes for localization without the need for additional infrastructure. However, it is sometimes necessary to install more anchors to improve accuracy, which increases the cost. Localization techniques based on RSSI measurements are the most common (Xie et al. 2014; Yang et al. 2020). However, other techniques use the measurement of channel state information (CSI) (Zhao et al. 2017) provided by some Wi-Fi boards. ToA has also been proposed, although it requires hardware modifications (Youssef et al. 2006). AoA localization techniques have recently been investigated in Kandel and Yu (2019) and Tian et al. (2019). Fingerprinting is a widely used indoor positioning method that is used with several wireless access technologies, including Wi-Fi, BLE and Zigbee (Nessa et al. 2020). This method can improve accuracy in NLoS environments, but it requires the creation of a database from training measurements, and the results depend on the density of reference points.

Technology	Coverage	Consumption	Accuracy	Cost	Method	Advantages	Disadvantages
UWB	30 m	Moderate	cm-m	High	ToA Fingerprint	Interference immunity	Dedicated hardware
Bluetooth	10-20 m	Low	m	Low	RSSI iBeacon	Widely available. Integrated in smartphones	Sensitive to multipath
WiFi	50 m	High	dm-m	Low	RSSI CSI AoA Fingerprint	Widely available	Sensitive to multipath. Complex processing
ZigBee	100 m	Moderate	m	Moderate	RSSI Fingerprint	Easy deployment	Sensitive to multipath. Dedicated hardware
Active RFID	20 m (indoor)	Low	< 1 m	Moderate	RSSI Fingerprint	Long battery life	Expensive readers. Property tags
Passive RFID	cm - few m	Very low	< 1 m	Low	RSSI Fingerprint	Tag low cost. No battery	Expensive UHF readers. Short range for HF system
Ultrasound	Couple of m	Low - Moderate	1.5 - 10 cm	Low	ToA	Widely used for distance measurements	Environmental noise. Air temperature and humidity dependence
Acoustic	Couple of m	Low - Moderate	cm	Moderate	ToA AoA	Smartphones compatibility	Ambient noise dependence. Requires mic. arrays & loudspeakers
Infrared	Few meters	Room-level	dm	Low	ToA TDoA	Cheap. Widely adopted for human detection	Fluorescent & sunlight interference. Required LoS
LIDAR	10 m	Very high	mm - cm	High	ToA AoA	Commercially available. No RF interference	Required LoS. Not wearable
Visual	10 m	High	mm - cm	Medium	AoA TDoA RSSI	Commercially available. No RF interference	Light dependence. Requires LoS
This work	<100 m	Very low	Room-level	Low	RSSI Fingerprint	Low number of anchors. Wearable. Interference immunity	Requires a database

Table 4.5. *Comparison of the reflection amplifiers proposed in the literature*

– *ZigBee-based localization:* Zigbee (IEEE 802.15.4 standard)-based localization techniques (Blumenthal et al. 2007; Pradityo et al. 2017) share the primary drawback of Wi-Fi-based techniques. But the use of Zigbee networks (mainly used for wireless sensor networks) is not as widespread in indoor environments as Wi-Fi networks, and requires an additional deployment that increases the cost of the localization system.

– *UWB-based localization:* UWB is one of the wireless solutions that has been the focus of attention for indoor localization. This interest arises from the fact that the use of large bandwidth yields high time resolution because it is possible to separate multiple reflections from the pulse. Consequently, UWB systems can obtain centimeter-level ranging accuracy under LoS conditions, even in indoor environments with multipath interferences (Dardari et al. 2009). The position is often determined by trilateration from the measured ToA from different anchors distributed in the coverage area. Localization accuracy depends on the number of anchors, their location and the LoS coverage. To address the problem of cost-effective UWB-based localization in complex indoor environments, hybrid trilateration and fingerprint algorithms have been developed (Djosic et al. 2021). The cost of the system depends on the number of anchors. UWB technology is not as extensive as other wireless technologies, and the cost of UWB transceivers is still expensive compared to other wireless systems.

– *Bluetooth-based localization:* the latest version of Bluetooth (IEEE 802.15.1) is known as Bluetooth low energy (BLE), which achieves a coverage range of 10–20 meters indoors with better energy efficiency compared to previous versions. The RSSI measurement provided by BLE devices can be used for localization (Faragher and Harle 2015). However, two BLE-based protocols, iBeacons (by Apple Inc.) and Eddystone (by Google, Inc.), have been designed for context-aware proximity services. From the messages received from the beacon and from the RSSI, users are classified into three range regions (immediate, near, far or unknown) (Andrushchak et al. 2018). Based on the user's proximity to any beacon, the mobile application triggers an action (e.g. sends a notification or a discount coupon) (He et al. 2015). Fingerprint approaches based on iBeacon have been proposed in the literature (Zuo et al. 2018; Dinh et al. 2020). Recently, the private company U-blox released the XPLR-AOA BLE and Bluetooth 5.1 kits for indoor positioning based on the AoA technique (XPLR-AOA-1 kit: Direction finding explorer kit with NINA-B4 Bluetooth 5.1 module n.d.).

– *RFID-based localization:* RFID systems can be classified into active and passive. Active RFID is often used to track assets outdoors, where great position accuracy (order of meters) is not required and when measurements are taken over long periods of time that require long-lasting batteries. They are based on UHF (ISM 433 or 868 MHz bands) and ISM 2.45 GHz. Essentially, two different types of active RFID tags are available: transponders and beacons. With transponders, communication works by means of backscattering, but using battery-assisted (BAP) devices. Active RFID transponders are commonly used in secure access control and payment systems in toll booths. Battery-assisted UHF tags that comply with Gen 2 standards (Jung et al. 2009) (and therefore, can be read with standard UHF readers) have recently been released which can extend the range of passive RFID systems up to 20 m. In these BAP tags, the uplink read range is limited by the sensitivity of the reader as in passive UHF tags. However, some active RFID tags use beacon tags. These tags use a transmitter that sends messages every few seconds. The use of transmitters extends the read range but can noticeably reduce the battery life depending on the transmission duty-cycle used. Active tag beacons are very common in the oil and gas industry, and for cargo tracking systems. For example, commercial Tagsense tags use a modified Zigbee transmitter, and Omni-ID integrates BLE and LoRaWAN transceivers. The read range in the open field is about 80 m depending on the antennas used (Cassel et al. 2017). This read range is significantly reduced in indoor environments to 20 m (Xu et al. 2018) due to multipath propagation. The sensitivity of a typical reader is in the order of -100 dBm. The price of active RFID tags is between $20 and $100 depending on their functionality and whether they must withstand harsh conditions. Although active RFID is designed for outdoor applications, it has also been explored for indoor localization applications. LANDMARC is one of the most popular indoor localization technologies used in active RFID tags (Ni et al. 2003; Jin et al. 2006; Wang et al. 2019). Passive tags, on the other hand, have shorter transmission ranges because they do not use a battery. They, therefore, have a lower read range than active RFID tags. UHF RFID can read up to a few meters, but HF RFID and NFC can read up to a few centimeters. Although they have the advantage of being low cost, UHF readers are more expensive than the NFC readers available in modern smartphones. An NFC-based navigation system has recently been proposed in Sakpere et al. (2017). The limitations in coverage make passive RFID suitable for proximity applications (Siachalou et al. 2019).

– *Acoustic-based localization:* radiofrequency (RF) signals share many characteristics with acoustic signals. Compared to more common RF signals, such as Wi-Fi, acoustic localization has been gaining strength because its main requirements are microphones and speakers, which are widely equipped in many smart devices (Cobos et al. 2017; Liu et al. 2020b). Another advantage of the acoustic signal is that the speed of sound is much lower than the speed of an RF signal, which is a positive factor in achieving higher accuracy. Different localization techniques are being used that combine multiple microphone signals, such as those based on ToA, TDoA, Doppler effect, direction of arrival (DOA) and steered response power (SRP) (Xie et al. 2014). An advantage of these devices is that they can use the microphones built into smartphones (Chen et al. 2016).

– *Ultrasound-based localization:* ultrasound-based localization technology uses ultrasound waves (typically at 44 kHz) to measure the ToA between a transmitter and a receiver (Ijaz et al. 2013; Medina et al. 2013). Due to the lower velocity of sound compared to that of an electromagnetic wave, an accuracy of centimeters can be achieved in range measurements. For node synchronization, the ultrasound localization system requires a secondary RF link. One disadvantage of these systems is the dependence of the sound velocity on the temperature and humidity and the high levels of environmental noise, which reduce accuracy and coverage.

– *Visual-based localization:* visual localization systems are commercially available on the market (e.g. Kinect (Shotton et al. 2011) or Wii (Schlömer et al. 2008) for games). These systems are not susceptible to RF interference from other devices, but are affected by ambient lighting conditions and pose privacy issues. Another disadvantage of this technology is the high computational requirements that make it difficult to integrate it into smartphones or wearables. These systems also require LoS, which limits their coverage to only a few meters (Alkhawaja et al. 2019).

– *Visible light-based localization:* visible light-based localization techniques use light sensors to measure the position and direction of LED emitters that work something such as iBeacons. The localization is based on AoA methods (Kuo et al. 2014). The advantage of this technology is its low cost, which makes it possible to install the LEDs over large areas. However, the biggest drawback with this technology, as with other techniques based on light, is the need for LoS.

– *LIDAR-based localization:* LIDARs and infrared cameras are becoming more popular in robotics (Xu et al. 2019) as their accuracy is in the order of millimeters. LIDAR is fast and accurate and is gaining popularity among car makers due to its use in the development of autonomous cars (Wolcott and Eustice 2014). Although LIDAR systems are becoming increasingly common, such systems are still expensive.

– *Infrared-based (IR) localization:* the low cost of IR positioning systems makes them a potential candidate at room level, although their range of coverage and accuracy are limited (Aitenbichler and Muhlhauser 2003). Technical limitations such as the required LoS between the transmitter and receiver and the interference of IR waves with fluorescent light and sunlight reduce the widespread usability of these systems. These issues have led researchers to explore alternative approaches to infrared-based indoor positioning.

As already addressed in the introduction of this chapter, the main novelty of the system proposed here is the utilization of a LoRa backscatter for localization. In addition to RFID systems, which also rely on backscattering communication, other wireless systems use direct communication between a transmitter and a receiver for localization. Consequently, the power consumption of the proposed system is lower than even BLE and Zigbee devices. A low-cost (< 5$) proof-of-concept of a wearable device has been presented. The prototype can run continuously for 24 days or for more than two years if the refresh rate is reduced to 1/10 seconds with a 500 mAh LiPo battery. The computational cost of the algorithm is low, and it is implemented on a server. Low power consumption is a key advantage, as it eliminates the need for frequent battery replacement, which can be an issue in ADL applications for use with elderly people. This long battery life also means that the system can be used to track assets in indoor environments. The high sensitivity of LoRa receivers and their great immunity to interferences (Elshabrawy and Robert 2018) due to the robustness of the CSS-modulation scheme allow a greater coverage range than other backscatters, including active RFID based on proprietary protocols. The use of lower frequencies reduces the undesirable effects associated with attenuation and diffraction compared to systems operating in the 2.45 GHz bands. In addition, in some countries, the ISM 433 MHz band can be used for LoRa (e.g. based on Semtech SX1278 (Lazaro et al. 2020b)), allowing the coverage range to be expanded. The result is better for indoor coverage, reducing the number of anchors or reference nodes and improving performance in wearable applications. LoRa is also a low-cost and low-power standardized wireless system that is continuously growing. Therefore, the cost of LoRa transceivers is lower than that of active UHF RFID readers. Low-cost Wi-Fi nodes with LoRa transceivers are used as receivers; therefore, the use of expensive multichannel gateways or routers is not required. Therefore, the overall cost of the system is lower than that of UWB and comparable to that of Wi-Fi or Zigbee-based systems. Device-free localization (DFL) techniques have been proposed using the variation of the RSSI received from different signals in a wireless network to detect and count the presence of human activity (Patwari and Wilson 2010). Although DFL has several interesting applications, the results are likely to be more sensitive to environmental changes. Fingerprint-based algorithms are often used in DFL. However, the training data require several measurements with a person standing at different positions to avoid the influence of measurement noise. In addition, training environments with multiple people that backscatter signals are challenging (Patwari and Wilson 2010). In the proposed system, the multiple signals backscattered on different objects in the scene are filtered from those produced by the LoRa backscatter because they change the frequency to another channel.

4.6. Conclusion

Techniques for estimating the absolute position of users in a coordinate system have been widely researched. Today, applications can locate a person or an object within a building by simply monitoring the room in which that individual or item is located. This chapter presents an indoor localization system with room-level accuracy and a focus on easy setup and a low-cost network. The system is based on the analysis of the backscattered signal strength received by LoRa transceivers deployed in an indoor environment. The backscatter device stands out for its low power consumption and long range. This chapter has explored different methods, ranging from a simple technique that assigns the room to the receiver with the highest measurement value to other more sophisticated techniques based on machine learning. The method based on the highest received signal strength exhibits great accuracy as long as the coverage is good in all rooms. Otherwise, to obtain better accuracy, methods based on machine learning are required. In these cases, a process to train the classification algorithms is necessary. This is a quick and straightforward process that only requires taking random samples in each room, without the need to determine the absolute coordinates at the sampling points. The proposed method also does not require the establishment of a propagation model, unlike other existing methods in the literature. The use of a low-cost, low-consumption LoRa backscatter device eliminates the need for frequent battery recharging, unlike in devices with higher consumption. Experimental results show that LoRa backscatter can be detected through walls in indoor environments. The LoRa backscatter can be easily integrated into a comfortable wristband. Due to the high sensitivity of the LoRa system, a large number of anchors are not required. The simulation and experimental results show that only one receiver in each room and a transmitter for every three to four receivers are needed to locate a person at the floor level in a standard home. The required wireless network can be implemented using low-cost modules available on the market without any modifications. The backend system can be implemented on any computer without the need for expensive servers using the MQTT protocol.

5

Backscattering for Implanted Medical Applications

This chapter assesses the feasibility of an implantable tag using LoRa backscattering technology for wireless communication. The proposed system is applicable to a broad number of medical use cases and can likewise be extended to environments characterized by severe signal attenuation, such as underwater or subterranean sensor deployments.

5.1. Introduction

The performance of wearable and implantable devices is often limited by the lifetime of batteries. A considerable part of the power budget of these devices is spent by the radiofrequency circuits (oscillators, receivers and transmitters) on wireless communication (Thomas et al. 2012). Recently, backscatter communications have raised great interest in a variety of applications, including implanted devices (Besnoff and Reynolds 2012; Iyer et al. 2016; Vasisht et al. 2018; Khaleghi and Balasingham 2019).

In this chapter, LoRa backscattering wireless communication is evaluated for the design of a low-power deep-implanted device operating in the Medical Implant Communication Service (MICS) bands and the ISM 433 MHz band. The proposed system implements the low-data rate LoRa packet OOK modulation described in Chapter 3. Therefore, commercial LoRa transceivers can be used for tag information demodulation. Backscatter communications based on LoRa wireless technology are suitable for communications with deep implants, as the strong attenuation of the

For a color version of all the figures in this chapter, see www.iste.co.uk/lazaromarti/low-power.zip.

body is compensated for by the sensitivity of the LoRa receivers. Furthermore, the high sensitivity of LoRa receivers can cope with the decrease of the antenna radiation efficiency due to the implanted tag miniaturization. Tag design is carried out using commercial off-the-shelf components, and the antenna is tuned at the desired frequency with a matching network and taking advantage of the frequency shift produced by the dielectric permittivity of the muscle and the tag coating.

The chapter is organized as follows. Section 5.2 describes the system overview with the phantom design, tag miniaturization, antenna tuning and link budget calculation. Section 5.3 presents the experimental results with the proposed system. Section 5.4 compares the system with other approaches reported in the literature. Finally, section 5.5 draws conclusions.

5.2. System design

5.2.1. *System description*

The system consists of a very little backscatter device implementing the already mentioned LoRa packet OOK modulation. For the transceivers, two LoRa32 boards have been used, one to transmit and the other to receive the backscattered packets in the adjacent channel. The proposed implanted device works in the MICS band (Islam and Yuce 2016), ranging from 401 to 406 MHz. Therefore, the Asian version of the commercial LoRa32 board has been used. In contrast to the European version that integrates an SX1276 operating at 868 MHz, the Asian version integrates an SX1278, which is capable of operating between 137 MHz and 525 MHz, making it suitable to operate in the MICS and in the Industrial Scientific and Medical (ISM) 433 MHz band, which is also frequently used for implanted medical devices (IMD). Alternatively, more expensive LoRa gateways integrate both transmission and multi-channel reception in the same device, making the system deployment more simple and allowing multiple tags to transmit packets to different channels simultaneously.

The OOK modulation reduces the cost of the implementation and is more than enough to send the information of the tag. However, if higher data rates are required, it can be further improved by demodulating the tag transmission by means of a software-defined radio. The technique described in Chapter 3 to maximize the power backscattered by creating two antipodal reflection coefficients is used for the implant design. Therefore, the tag antenna is loaded with an open-circuit and a short-circuit using the ADG902 switch. The switch is controlled by a low-power square wave LTC6907 oscillator from Analog Devices (Linear Technologies), which can be adjusted with a simple resistor for an output frequency between 40 kHz and 4 MHz. The power consumption of the oscillator is 36 μA for an output frequency of 400 kHz. Moreover, the LTC6907 has a 1% frequency accuracy from $0°C$ to

$70°C$, which is enough for the LoRa-packet OOK modulation. The tag controls the oscillator by means of the low-power AtTiny402 microcontroller. For the antenna, a low-profile ceramic antenna from Johanson Technology designed to operate at the 868–900 MHz band is used. However, with some adjustments (and under specific implant conditions), it can be successfully adjusted in the MICS and 433 MHz ISM bands. Figure 5.1 shows an image of the prototype, and Figure 5.2 shows a spectrogram with the tag successfully modulating the LoRa packets at the 406 MHz band.

Figure 5.1. *Photograph of the LoRa backscatter. The bottom layer with the microcontroller and the top layer with the RF switch, antenna and oscillator*

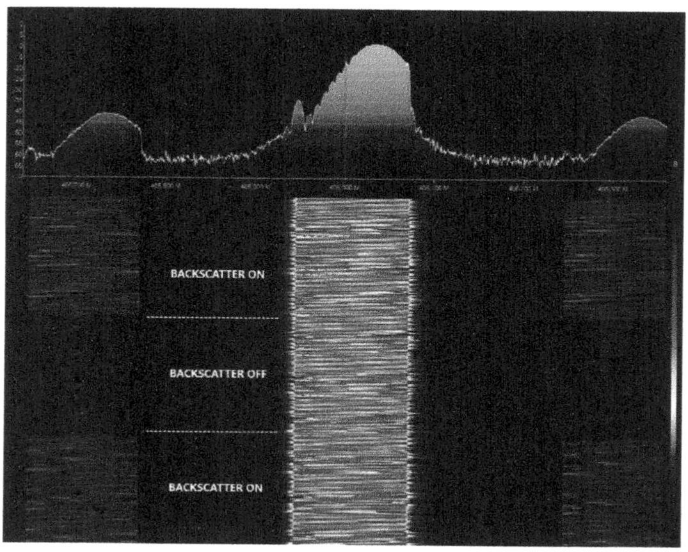

Figure 5.2. *Measured spectrogram with the backscatter enabled and disabled for SF = 12, BW = 125 kHz*

As previously described in Chapter 3, the backscattered power is maximized if the antenna resonates in the band ($[Im]Z_a = 0$) and the two reflection coefficients have a phase shift of 180° to one another. Reflection coefficients at each state (open-circuit and short-circuit) are measured as a function of the frequency using the setup depicted in Figure 3.2 of Chapter 3. Markers at the 406 and 433 MHz bands show the insertion loss introduced by the ADG902 switch. A loss of 1.6 dB with respect to the ideal short-circuit and open-circuit cases is observed. The measured phase difference between both states is close to 180°.

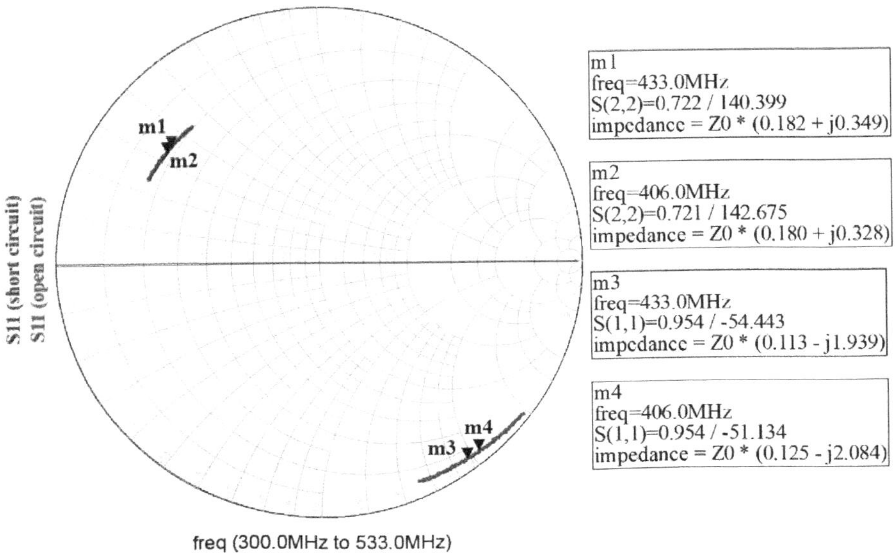

Figure 5.3. *Measured reflection coefficients of the ADG902 switch (S_{11} open-circuit state (red) and short-circuit state (blue)) at the MICS (406 MHz) and 433 MHz ISM bands*

5.2.2. *Phantom design*

In order to develop an antenna for a deep implant, a phantom material needs to be designed to simulate the muscle tissue within the desired frequency range. The dielectric properties of different tissues have been reported in the literature (Gabriel et al. 1996). Consequently, several phantom materials have been proposed for various frequency ranges and applications (Chou et al. 1984; Peterson et al. 2010; Castelló-Palacios et al. 2017, 2019). In this particular case, the dielectric properties must be designed to account for the shift in the antenna's frequency response caused by the high dielectric constant of the tissues, as well as to simulate the attenuation of the body and the reduction in the antenna's radiation efficiency due to the high tissue conductivity. The goal is to find a simple phantom for the ISM band (400–433 MHz). According to data provided by Gabriel et al. (1996), the dielectric constant of muscle is approximately 57, and its conductivity is 0.8 S/m at 406 MHz. For frequencies greater than 100 MHz, the muscle dielectric constant is lower than that of water (Chou et al. 1984), making a phantom based on a saline water solution more suitable, as antennas can be used more easily within it compared to a solid phantom. In addition, manufacturing such a phantom is easier and cheaper than manufacturing solid phantoms. In Castelló-Palacios et al. (2017), a mixture of water with NaCl and sucrose was proposed as a phantom design at 2.45 GHz. The current study follows a similar procedure but adjusts the parameters for the 400–500 MHz band. Permittivity is measured using an open-ended coaxial probe (Burdette et al. 1980) from Keysight (slim form probe 85070E/N1501A model) with a vector network analyzer (VNA) model Keysight PNA E8364C. Figure 5.4(a) displays the real and imaginary parts of the permittivity as a function of the mass fraction (percentage by weight) of sugar at 406 MHz. The imaginary part remains nearly constant, whereas the real part decreases with the increase of sugar. It can be observed that for a sugar percentage of approximately 45%, the real part closely matches the value for muscle. Figure 5.4(b) shows the complex permittivity for a saline water solution with 45% sugar, varying the quantity of NaCl at 406 MHz. Thus, by adding NaCl, the conductivity increases, and by adding sugar, the real part is reduced. A solution consisting of 45% sucrose, 0.8% NaCl and water (percentages are by weight) is selected. Figure 5.5 compares the data reported by Gabriel et al. (1996) with the measurements obtained here at ambient temperature (25°C). The agreement is good within the 400–600 MHz band. The power transported by the wave decreases exponentially following Lambert's law. The attenuation constant introduced by the tissue can be computed using (Ayappa et al. 1991):

$$\alpha\left(\frac{dB}{m}\right) = 8.686 \cdot 2\pi f \sqrt{\frac{\mu_0 \varepsilon_0 \varepsilon'}{2}[\sqrt{1+(\frac{\varepsilon''}{\varepsilon'})^2}+1]} \qquad [5.1]$$

where $\varepsilon_c = \varepsilon' - j\varepsilon''$ is the complex relative permittivity, ε_0 and μ_0 are the vacuum permittivity and permeability, respectively, and f is the frequency. Considering the muscle complex permittivity, a typical attenuation of 165 dB/m or a power penetration depth of $d_p = 1/(2\alpha(N_p/m)) = 2.6$ cm is introduced by the muscle at 406 MHz. Equation [5.1] has been derived for incident plane waves in large bodies without considering reflections within the materials at the opposite interface with the air. Ayappa et al. have shown that the above relationship only applies if the thickness of the tissues is larger than the critical length, which is about 5.4 d_p (Ayappa et al. 1991). Otherwise, computer simulation models are required to predict more realistic power distributions.

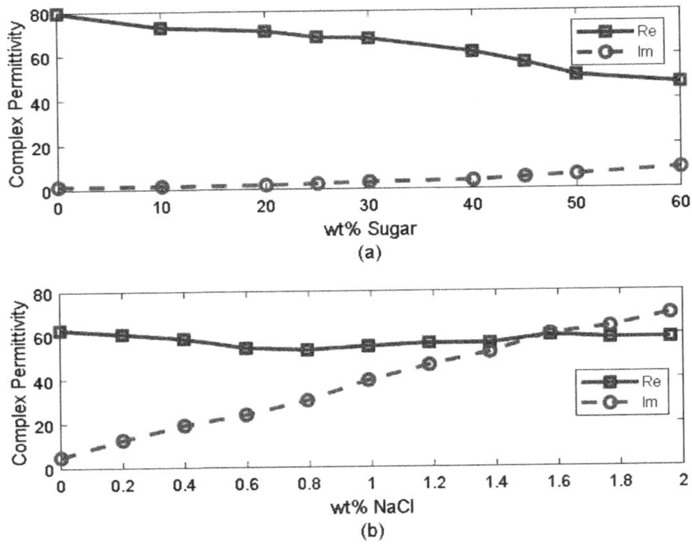

Figure 5.4. *(a) Measured dielectric permittivity (real and imaginary parts) as a function of sugar mass fraction at 406 MHz and (b) measured dielectric permittivity (real and imaginary parts) as a function of NaCl mass fraction and 45% sugar at 406 MHz*

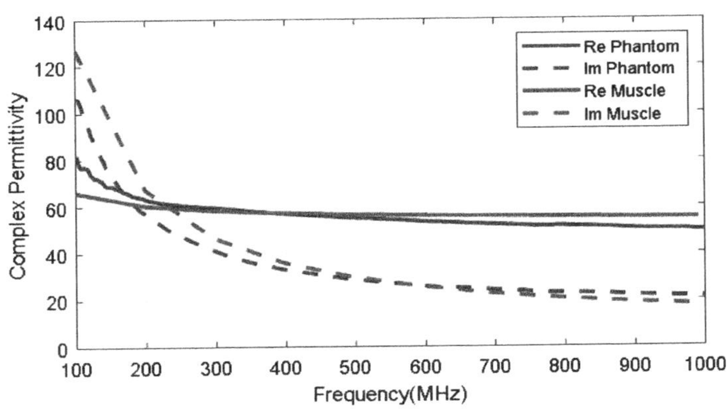

Figure 5.5. *Measured dielectric permittivity (real and imaginary parts) as a function of frequency compared to muscle tissue data (Gabriel et al. 1996)*

5.2.3. *Antenna for deep implanted backscatter*

Several implanted antennas have been proposed in the literature for ISM bands and MICS (Soontornpipit et al. 2004; Murphy et al. 2011; Kiourti and Nikita 2012). A ceramic (alumina) antenna with a superstrate was proposed in Soontornpipit et al. (2004). Chip antennas made with low-temperature co-fired ceramics (LTCC) are available on the market and are extensively used in wireless modules (ISM band, Bluetooth, GPS, WLAN, etc.). The high permittivity of these materials enables the antennas to be smaller so that they can be integrated into small circuits. In the present study, a miniature commercial LTCC ceramic antenna (0900AT43A0070, Johanson Technology, Camarillo, CA, USA) designed for operation in the air at 900 MHz is used to operate as an implant in the 402–433 MHz band. The antenna is 7.7 mm long, 2 mm wide and 0.8 mm high. Simulations and experiments are conducted to study the feasibility of commercial ceramic antennas for devices implanted in MICS and the 433 MHz band. The small size of the antenna (0.0715λ in the body) means that it is electrically small compared with the wavelength, so neither the pattern diagram nor the directivity will be greatly affected, and it has an omni-directional radiation pattern. Due to the high permittivity surrounding the antenna, the effective electrical length of the monopole decreases and changes the antenna impedance. This drawback can be overcome with a matching network that has discrete components (inductors and capacitors) as that proposed by the manufacturer for air but tuned to the impedance of the implanted antenna. However, the main drawback is that the antenna radiation efficiency is reduced by the lossy tissue material close to the antenna. This reduction in radiation efficiency, in turn, reduces the antenna gain, and consequently, the $\Delta\sigma$ and the backscattering power are also reduced.

IMD must be covered by a biocompatible material to prevent contamination (Donaldson and Sayer 1981) and metallic oxidation, and the possibility of a short-circuit due to the high conductivity of the tissues (Merli et al. 2010). The coating material acts as a superstrate, and the thickness and the material have a considerable influence on the radiation efficiency (Soontornpipit et al. 2004). Increasing the thickness of this coating material or the air gap between the cover and the antenna improves the radiation efficiency (Merli et al. 2010) as well as increases the size of the implanted device. Therefore, there is a constraint between radiation efficiency and size. In the present study, the coating is implemented using rubber silicone ($\varepsilon_r = 2.9$, $\tan\delta = 0.02$) (Teo et al. 2016). Figure 5.6b shows the reflection coefficient of the antenna connected to a short transmission line (width 1.85 mm) printed on a 32 mil Roger RO4003C substrate, within the body with a 1 mm thick coating of silicone. The size of the ground plane is 10 mm × 10 mm. An image of the 3D simulated model is shown in Figure 5.6a. This figure shows that the antenna is completely mismatched in the 400–433 MHz band. Therefore, a matching network is required. Assuming a well-matched design, the antenna gain should be close to the simulated antenna gain shown in Figure 5.6b. Figure 5.7 shows the gain, directivity and radiation efficiency at 406 MHz as a function of the thickness of the silicone coating. The simulations have been performed with the full-wave Ansys HFSS simulator using the model provided by the antenna manufacturer. A box of muscle tissue ($\varepsilon_r = 57$, $\sigma = 0.8$ S/m) surrounding the antenna is used to simulate the effect of the body. The antenna directivity is hardly affected by the thickness of the coating, but the radiation efficiency, and therefore the gain, increases with thickness. The efficiency fluctuates between 9.5% and 16%, much lower than that obtained in air (approximately 27%). Figure 5.8 shows the effect of the coating permittivity. This figure assumes that the coating material is 1 mm thick. Radiation efficiency increases slightly as the permittivity of the coating increases. However, it should be noted that coatings based on ceramic substrates with high permittivity are difficult to drill and cut.

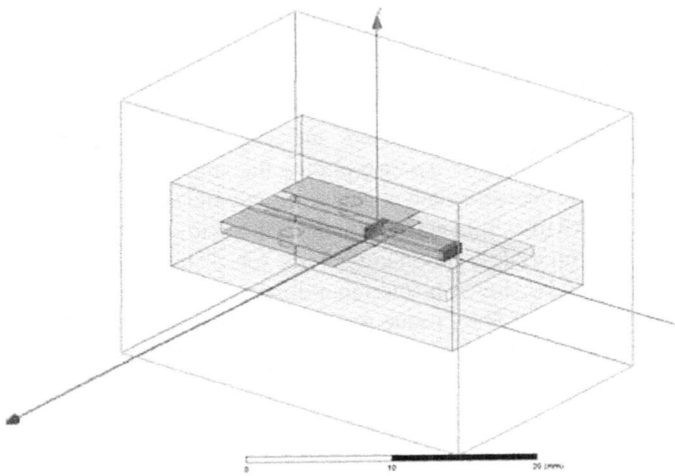

(a)

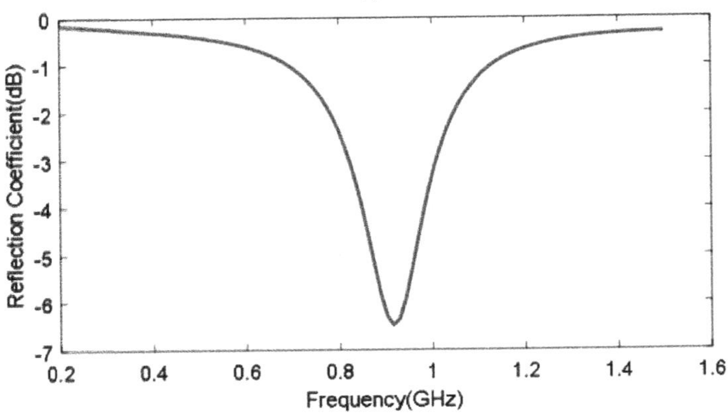

(b)

Figure 5.6. *(a) Image of the simulated antenna with HFSS, (b) simulated antenna reflection coefficient as a function of the frequency for the antenna with a 1-mm thick silicone coating*

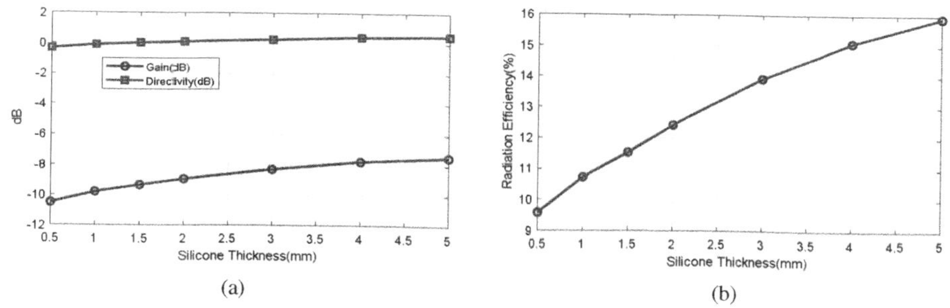

Figure 5.7. *(a) Simulated gain and directivity at $\theta = 0°$, $\phi = 0°$, and (b) radiation efficiency as a function of the silicone thickness at 406 MHz*

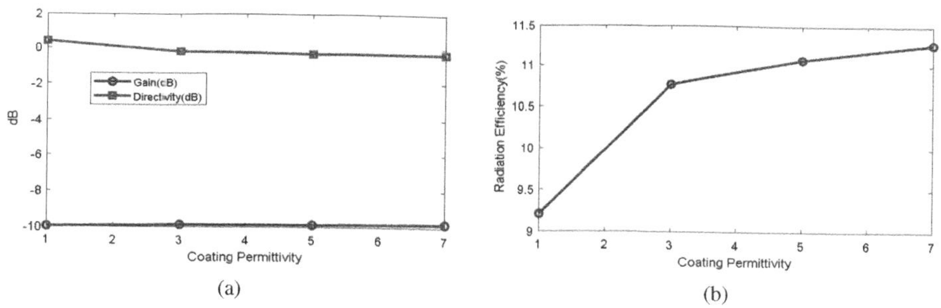

Figure 5.8. *(a) Simulated gain and directivity at $\theta = 0°$, $\phi = 0°$ and (b) radiation efficiency as a function of the coating permittivity (1-mm thick coating thickness) at 406 MHz*

In order to design the matching network, a prototype with an SMA connector has been manufactured (Figure 5.9a). The size of the board (7.5 mm × 21 mm), ground plane (7.5 mm × 12 mm) and the position of the discrete matching components is the same as in the implanted prototype (Figure 5.1). The impedance of the antenna is measured with a VNA connecting a thru line (zero Ohm SMD resistance), after which the reference plane has been shifted to obtain the antenna impedance. Then, an LC matching network is designed using the Smith chart. Figure 5.9b shows the measured reflection coefficient of the antenna in the phantom material before and after the matching network. The frequency band can be tuned by modifying the components of the matching network. A 56 nH inductance in series and a 0.5 pF in parallel are used to match the antenna at 406 MHz. The measured reflection coefficient is below -10 dB between 387.28 MHz and 469.3 MHz (82.1 MHz bandwidth or 19.2%), ensuring operation in both 433 MHz and 406 MHz bands.

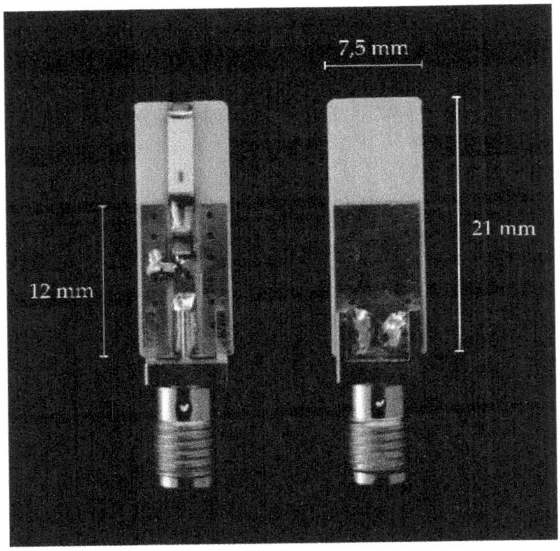

(a)

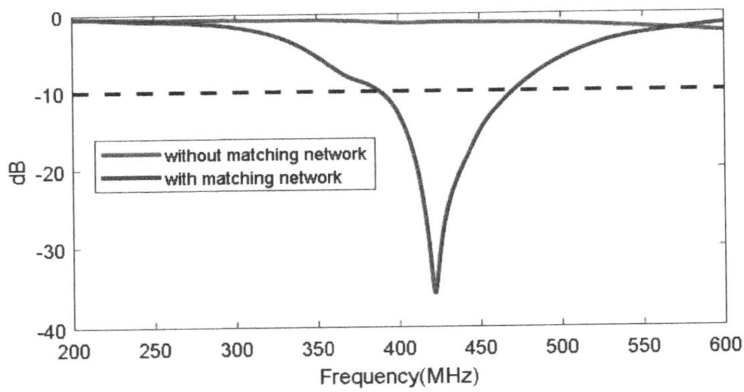

(b)

Figure 5.9. *(a) Photography of the prototype manufactured for the antenna characterization; (b) measured reflection coefficient of the antenna in the phantom with and without the matching network*

Table 5.1 compares the designed antenna with other implanted antennas proposed in the literature for the MICS band (Liu et al. 2008; Huang et al. 2011; Kiourti et al. 2011; Nikolayev et al. 2017) using different miniaturization techniques. The miniature antenna proposed in the present study achieves a high gain and radiation efficiency, and a bandwidth that is moderate but enough to cover the MICS and 433 MHz bands.

It is also available commercially, which simplifies the commercialization of future implanted devices.

Reference	Volume (mm^3)	Antenna Type	Gain (dB)	Efficiency (%)	BW at -10 dB (MHz)
Huang et al. (2011)	245	PIFA with superstrate	-7	39	115
Kiourti et al. (2011)	32.7	Meander patches	-45	0.81	40
Liu et al. (2008)	190	Spiral patch	-26	0.61	50
Nikolayev et al. (2017)	705	Alumina capsule $\lambda/2$ SIR	-22	0.4	16
This study	28 (236)[1]	Ceramic SMD	-9.5	10.8	82.1

Table 5.1. *Comparison with other antennas in the Medical Implant Communication Service (MICS) band reported in the literature*

5.2.4. *Link budget*

In this section, the propagation model presented in section 4.2.3 of Chapter 4 is used to calculate the link budget based on the system parameters. The objective is to determine the maximum depth and coverage area at which the implanted backscatter can be read.

The received power can be calculated using the Friis transmission equation, already presented in section 4.2.3, but now considering the attenuation of the body α(dB/m), and the implant depth d. Therefore, equation [4.1] can be rewritten as:

$$P_R = \frac{P_T G_T}{4\pi d_T^2} \cdot \Delta\sigma \cdot 10^{-0.1\alpha \cdot 2d} \cdot \frac{1}{4\pi d_R^2} \cdot (\frac{\lambda^2}{4\pi}) G_R \qquad [5.2]$$

The parameters used in the simulations are listed in Table 5.2. The LoRa receiver sensitivity (S) can be determined using equation [4.4] from Chapter 4. It depends on the noise figure of the receiver, the bandwidth (BW) and the signal-to-noise ratio (SNR), which is a function of the spreading factor, as shown in Table 5.3 (Semtech Corporation 2020, 2013). LoRa SX1278 transceivers allow operating with one of the following predefined bandwidths: 7.8, 10.4, 15.6, 20.8, 31.5, 41.7, 62.5, 125, 250, 500 kHz. The calculation of the sensitivity in Table 5.3 assumes a typical BW of 125 kHz. Thus, the sensitivity decreases if a narrower bandwidth is used in accordance with equation [4.4] (12 dB for the lower bandwidth compared to 125 kHz). Table 5.3

1 Including a coating of 1 mm, a matching network and a ground plane.

also shows the time on air and the equivalent bit rate computed from the equations given in Semtech Corporation (2013, 2020), considering 10 bytes of payload with a preamble of 6 bytes and a CR coding rate of 4/5.

Parameter	Value	Unit
Transmission power P_T	20	dBm
Transmission antenna gain G_T	0	dB
Receiver antenna gain G_R	0	dB
Tag antenna gain G_a	-10	dB
Carrier frequency	406	MHz
Attenuation per unit length α	165	dB/m
Exponential decay factors (n_1, n_2)	2.5	
$\lvert\Delta\Gamma\rvert^2$	2.8[2]	
Modulation factor (m)	$1/\pi^2$	
Noise figure of the receiver (NF)	6	dB
Bandwidth (BW)	125	kHz

Table 5.2. *Parameters used to calculate the link budget*

Spreading factor SF	Chips/Symbol[3]	SNR (dB)	ToA of a 10-byte packet[4] (ms)	Bit rate (bps)	Sensitivity S (dBm) for BW = 125 kHz[5]
7	128	-7.5	56	5470	-124.5
8	256	-10	103	3125	-127.0
9	512	-12.5	205	1758	-129.5
10	1024	-15	371	977	-132.5
11	2048	-17.5	741	537	-134.5
12	4096	-20	1483	293	-137.0

Table 5.3. *Parameters as a function of the spreading factor*

2 From measurements shown in Figure 5.3.

3 Chips/Symbol = 2^{SF}.

4 With a preamble of six symbols and CR = 4/5.

5 S can be reduced 12 dB for BW = 7.8 kHz.

However, due to multipath propagation and reflections on the ground and objects in indoor environments, the power decays with exponential factors that are not those given in the Friis model considered in equation [5.2]. The received power or received signal strength indicator (RSSI) can be expressed using the empirical model [4.2] of section 4.2.3 without considering the losses introduced by the walls. This model is an extension to the backscatter channel of the empirical propagation model for indoor environments (Goldsmith 2005; Huang et al. 2011):

$$P_R(d_{TX}, d_{RX})(dBm) = P_R(d_0, d_0) - 10n_1 log(\frac{d_{TX}}{d_0}) - 10n_2 log(\frac{d_{RX}}{d_0}) + X \quad [5.3]$$

In this model, the received power is modeled as an average term and a random term X, where d_0 is a reference distance (e.g. the midpoint between the transmitter and receiver $d_0 = (d_{TX-RX}/2)$) and $P_R(d_0, d_0)$ is the average received power at this reference distance. The exponential decay factors n_1 and n_2 are a function of the environment and the antenna height but for indoor environments, they have values of about 2.5–3 (Lazaro et al. 2009; Tuset-Peiró et al. 2014). From the Friis equation, $P_R(d_0, d_0)$ is a function of the transmitted power and the antenna gains. However, it is also a function of the antenna height and the diffraction. $P_R(d_0, d_0)$ can be found experimentally by regressing the measured data. Nevertheless, in this section, simulations are performed considering the following expression derived from the Friis transmission equation [5.2]:

$$\begin{aligned} P_R(d_0, d_0)(dBm) &= P_T(dBm) + G_T(dB) + G_R(dB) - 20log(4\pi) \\ &+ 10log(\frac{\lambda^2}{4\pi}) + RCS_{dif}(dB) - \alpha(dB/m)2d \\ &- 10n_1 log(d_0) - 10n_2 log(d_0) + L_{obs}(dB) \end{aligned} \quad [5.4]$$

L_{obs} is a term in dB that accounts for the attenuation for diffraction that depends on the antenna height (Goldsmith 2005). The losses for ground diffraction can be neglected if the antenna height is higher than $0.6R_1$, where R_1 is the radius of the first Fresnel zone (Goldsmith 2005):

$$R_1 = \sqrt{\lambda \frac{d_{TX} d_{RX}}{d_{TX} + d_{RX}}} \quad [5.5]$$

For a distance between the transmitter and receiver (backscatter) of 2 m, and antennas of the same height, the minimum antenna height is 0.36 m at 406 MHz. Therefore, for a typical antenna height of approximately 1–2 m, it can be expected that the effect of ground diffraction will be small.

The random variable X takes into account the attenuation due to multipath propagation. It can be regarded as a log-normally distributed random variable

$X(dB) \sim N(0, \sigma)$, where σ in dB is the standard deviation and experimental results have given values between 2 and 4 dB for a single path (Tuset-Peiró et al. 2014) (in the case studied here, it is expected to be double as a result of the backscatter channel). The coverage probability ($Prob$), which we recall is the probability that the average received power or RSSI, $P_R(dBm)$, will be higher than the receiver sensitivity S is computed using equation [4.3] of Chapter 4.

Two cases are analyzed: monostatic and bistatic. In the monostatic case, the LoRa transmitter and receiver gateway are close together ($d_{TX} = d_{RX}$). Figure 5.10 compares the maximum depth or distance inside the body that the backscatter can communicate for SF = 7 and SF = 12 with BW = 125 kHz. Figure 5.11 shows the maximum depth as a function of the transmitter-to-body distance and receiver-to-body distance in the bistatic case for SF = 7 and SF = 12 with BW = 125 kHz, respectively. In these simulations, the exponential decay factors (n_1 and n_2) have been considered equal to 2.5. Depths can be greater than 10 cm at 4 m. In the bistatic case, depths can be greater if the transmitter is closer to the body than the receiver.

These simulations take into account a fading margin due to a multipath propagation of 10 dB. Results show the potential to read low-rate data from implants using frequency-shifted LoRa backscatter. The number of packets per unit of time increases if the spreading factor is low (e.g. SF = 7) as can be seen in Table 5.3 because the time on air is shorter than at high spreading factors. However, the sensitivity and robustness of the system are better when the spreading factor is high. In the bistatic case, to achieve better coverage, it is assumed that the transmitter will be located near the patient (e.g. close to the bed) and the receiver somewhere else in the room.

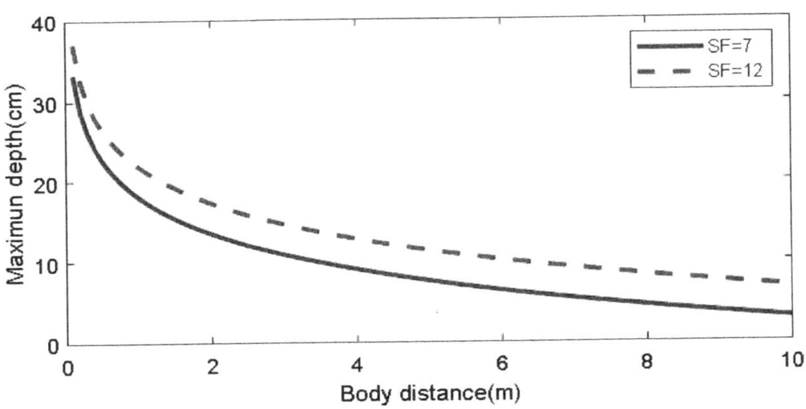

Figure 5.10. *Simulated maximum depth for an implanted backscatter as a function of the distance to the body using the parameters of Table 5.1 for the monostatic case with a spreading factor SF = 7 and 12. Propagation model parameters:* $n_1 = n_2 = 2.5$*, fade margin = 10 dB*

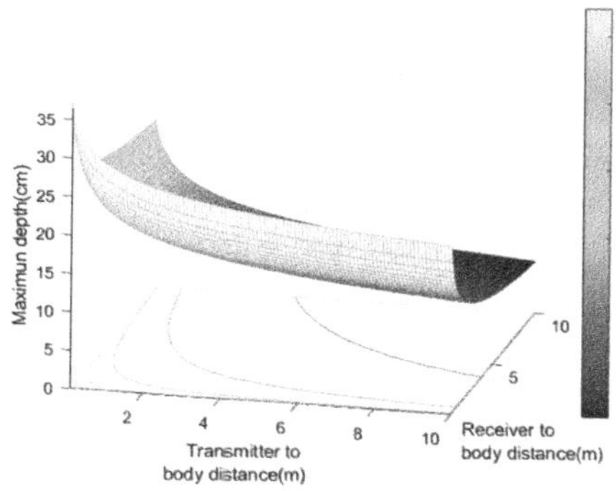

(a)

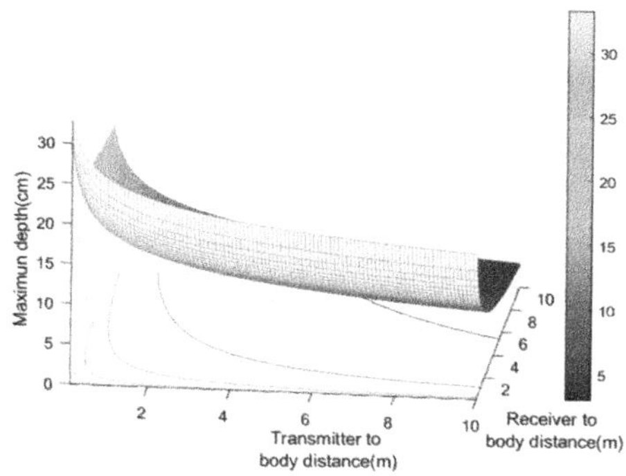

(b)

Figure 5.11. *Simulated maximum depth for an implanted backscatter as a function of the distance of the transmitter and receiver to the body using the parameters of Table 5.1 for the bistatic case with a spreading factor SF = 7 (a) and SF = 12 (b). Propagation model parameters:* $n_1 = n_2 = 2.5$*, fade margin = 10 dB*

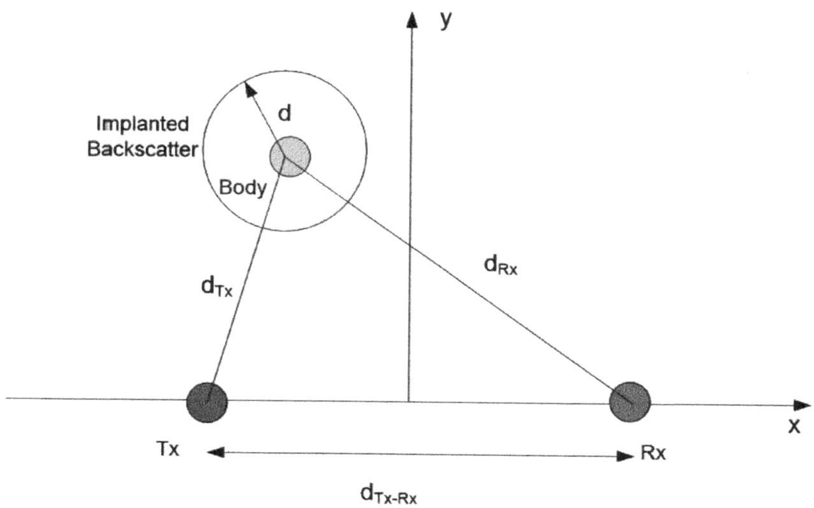

Figure 5.12. *Scheme used in the RSSI simulations*

A simplified scenario shown in Figure 5.12 consists of the transmitter, the receiver and the implanted backscatter on the same plane. All these elements are used to estimate the coverage from the proposed propagation model. In this scenario, it is assumed that the transceivers are fixed and the backscatter can move around different positions.

Figures 5.13 and 5.14 show the simulated coverage results for backscatter at a depth (d) of 5 cm and 10 cm, respectively. They depict the average received power computed with equation [5.3] and [5.4] considering the ideal free space exponential factors $n_1 = n_2 = 2$ and $n_1 = n_2 = 2.5$ as a function of the position of the backscatter. The monostatic and bistatic cases (with a distance between the transmitter and receiver $d_{TX-RX} = 1m$) are shown. For devices implanted on the surface or at small depths (<5 cm), the read range area can be about 4 m × 4 m, whereas for deep implanted devices (10 cm), a read range area of about 1.8 m × 1.8 m can be covered.

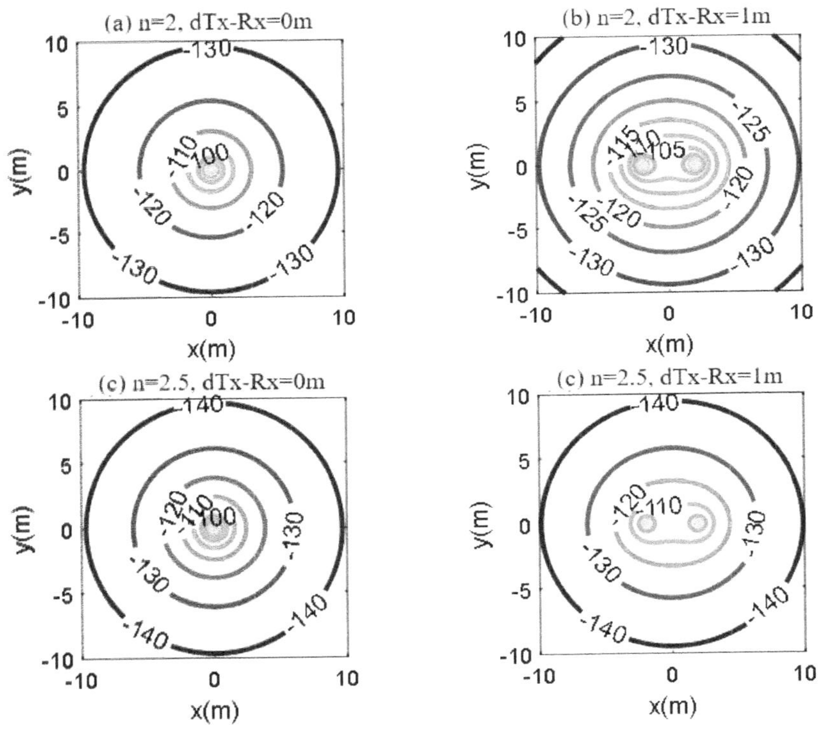

Figure 5.13. *Simulated RSSI as a function of the location of the implanted backscatter assuming a depth of 5 cm inside the body for the monostatic case ($d_{Tx-Rx} = 0$ m) and bistatic case ($d_{Tx-Rx} = 1m$). Path loss exponential factors $n_1 = n_2 = 2$ and $n_1 = n_2 = 2.5$*

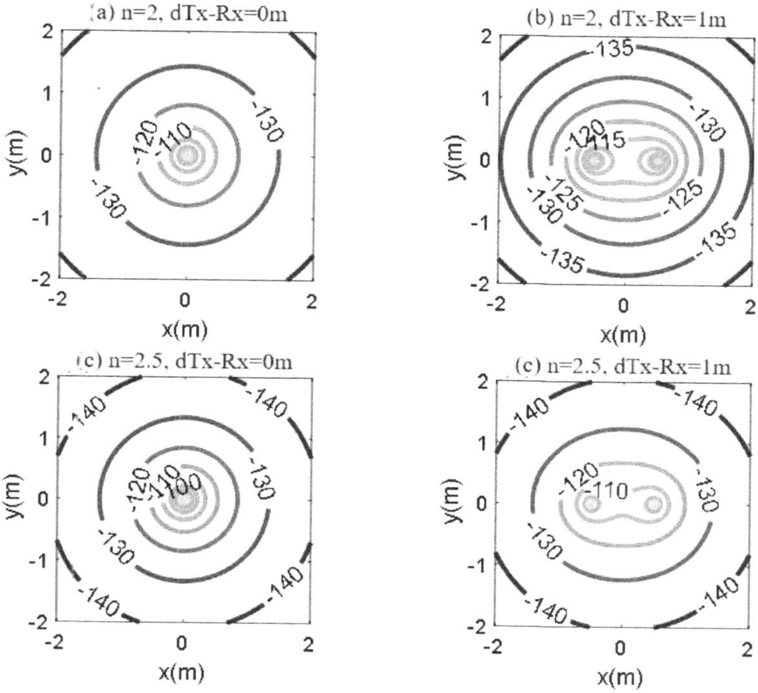

Figure 5.14. *Simulated RSSI as a function of the location of the implanted backscatter assuming a depth of 10 cm inside the body for the monostatic case ($d_{Tx-Rx} = 0$ m) and bistatic case ($d_{Tx-Rx} = 1$ m). Path loss exponential factors $n_1 = n_2 = 2$ and $n_1 = n_2 = 2.5$*

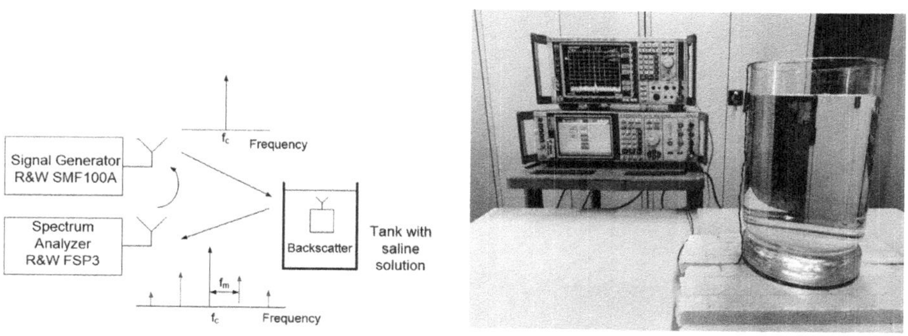

Figure 5.15. *Block diagram of the setup used for the characterization of the backscatter and photograph*

5.3. Results

A backscatter prototype has been designed using the load-modulation front-end described in Chapter 3. The prototype includes an antenna with a matching network, an RF switch (ADG902) (Microchip Atmel) used as a modulator and the low-power oscillator (LTC6907). The microcontroller generates the baseband signal with the test messages. The backscatter is covered with heat-shrink silicone (approximately 1 mm thick). This proof-of-concept prototype is externally biased with a battery pack connected by two wires that are also used to support the device. A photograph is shown in Figure 5.15. In order to check the correct modulation of the backscatter inside the body, it is inserted into a 20 cm diameter cylindrical vessel (see Figure 5.15). The vessel is filled with a mixture of saline solution with sugar described in the previous section. A signal generator is connected to a transmitting antenna, and a spectrum analyzer is used as a receiver. Two whip monopoles tuned at 406 MHz have been used in the generator and the spectrum analyzer. The distance from the Tx and Rx to the vessel is set to 1 m. The signal generator frequency is swept from 350 MHz to 450 MHz, and the backscattered power is received by the spectrum analyzer. Figure 5.16 shows the spectrum in which the 406 MHz carrier signal and the sidebands at the backscatter modulating frequency can be seen. The central peak at this carrier frequency is due to the coupling between the transmitter and receiver. The sideband peaks are spaced from the carrier at the same frequency at which the ADG902 is switching (300 kHz in this experiment). Figure 5.17 shows the measurement of the received power at the sideband as a function of the generator frequency. The maximum received power is observed as approximately 408 MHz. It can be noted that the backscattering implant covers the desired frequency band, and proves that the antenna is correctly tuned within the phantom.

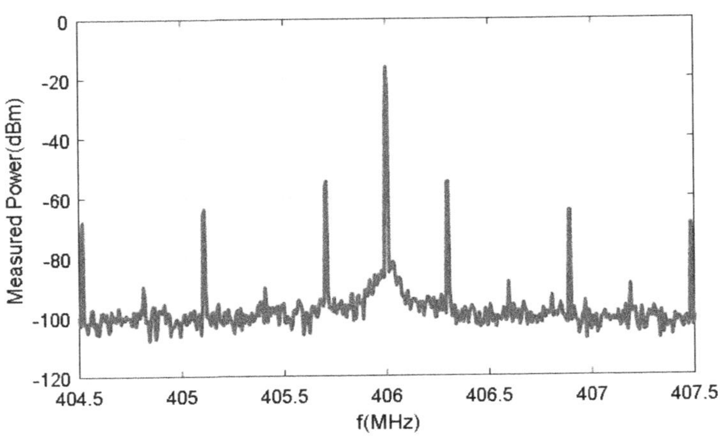

Figure 5.16. *Spectrum of the receiver signal when the backscatter is illuminated at 406 MHz using the setup in Figure 5.15*

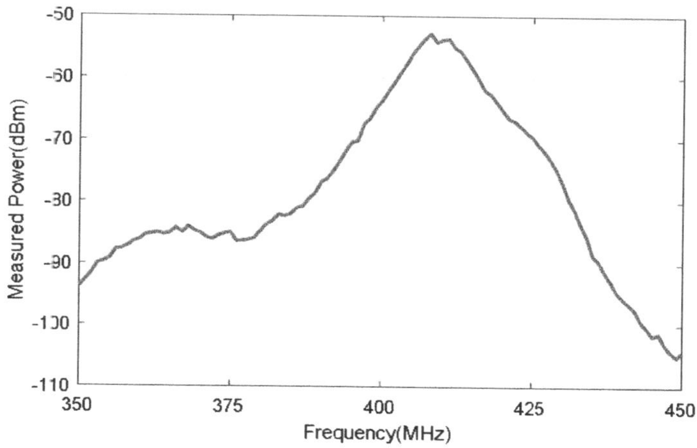

Figure 5.17. *First harmonic ($f_c + f_p$) measured power as a function of the frequency with the backscatter immersed in the vessel described in Figure 5.15*

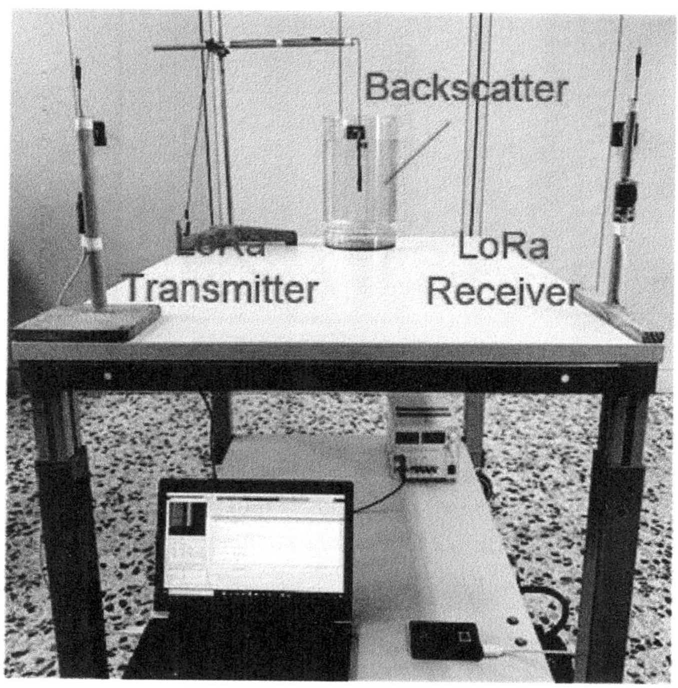

Figure 5.18. *Photograph of the experimental setup in the laboratory*

To test the reliability of the system, some experimental results have been carried out using two LoRa32 boards as a transmitter and receiver. The central frequency of the transmitter is 406 MHz, and the switching frequency is 300 kHz. Therefore, the receiver is tuned at 406.3 MHz. The LoRa is configured to transmit and receive packets with SF = 12 and BW = 125 kHz for better sensitivity. Figure 5.18 shows an image of the setup used in the experiments.

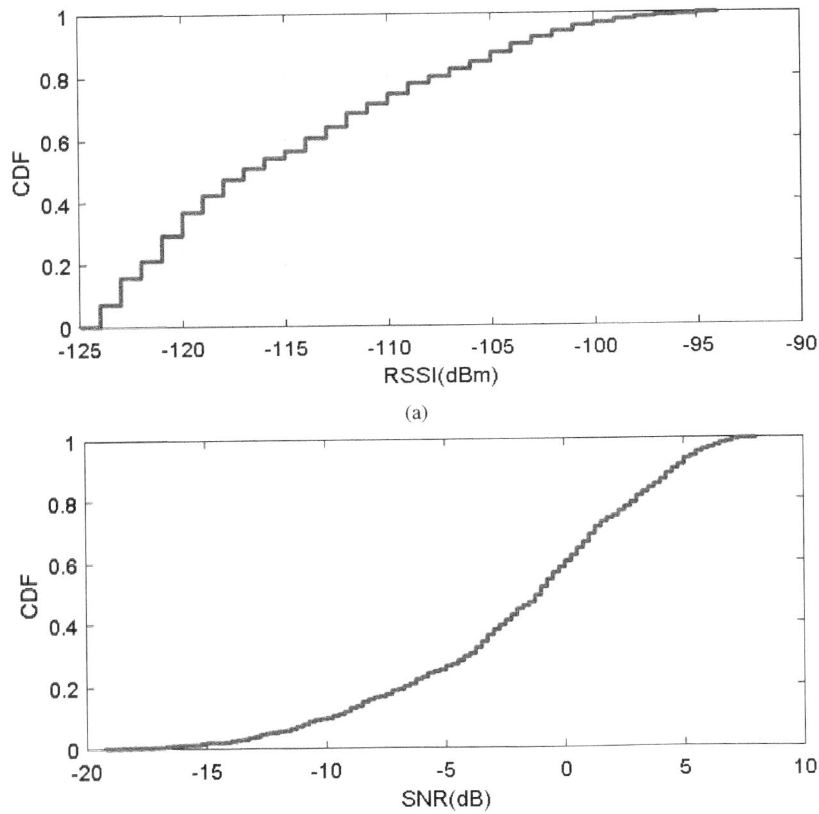

(b)

Figure 5.19. *Complementary distribution function of the measured RSSI (a) and SNR (b) with the backscatter immersed in the vessel for a transmitter located 80 cm from the vessel*

Figure 5.19 shows the complementary distribution function (CDF) plot of the RSSI and the SNR in the shifted channel measured by the LoRa receiver. In order to obtain the data, the receiver is tuned at the backscattered channel and located at random points (uniformly distributed) in the laboratory within a coverage area of 4×4 m.

The distance from the LoRa transmitter to the vessel is 80 *cm* at a height of 1 *m* from the ground. The backscatter is immersed in the center of the 20 *cm* diameter vessel with the phantom solution. It can be observed that the RSSI is typically higher than -123 *dBm* (less than 10% of the locations receive levels below -123 *dBm* and SNR below -15 *dB*). Therefore, most locations are above the sensitivity of the receiver, and the backscattered packets can be successfully received.

In order to show the coverage area, Figure 5.20 shows the heat maps of the measured RSSI for the transmitter located 80 *cm* from the vessel with the backscatter prototype immersed at a depth of 10 *cm* in the center of the vessel. The backscatter prototype and the transmitter are fixed, whereas the receiver is moved along different locations in the room (laboratory). Under these circumstances, data can be received within the laboratory (area 5 *m* × 10 *m*) and through walls provided the receiver is located in neighboring rooms on the same floor. This result is in accordance with Figure 5.11(b) which shows that devices implanted at a 10 *cm* depth can be detected up to 10 *m* from the receiver when the transmitter is close to the body (80 *cm* in this case).

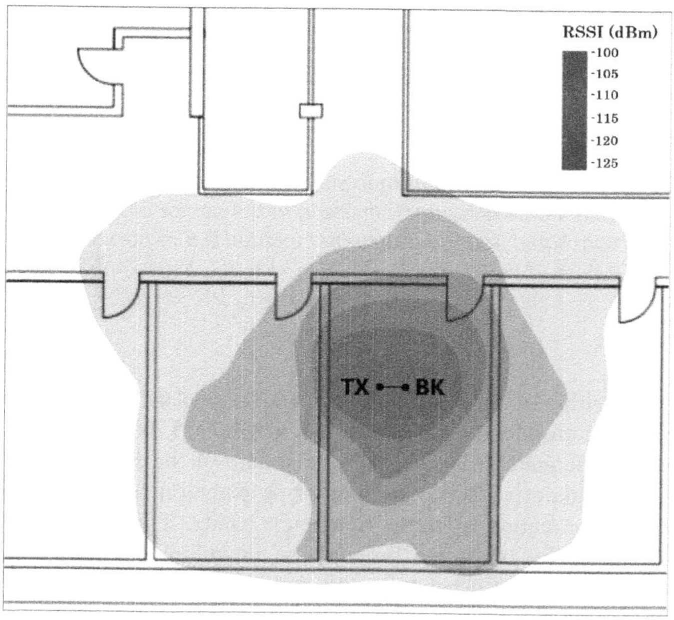

Figure 5.20. *Measured RSSI heat map for the transmitter located 0.8 m from the vessel with the backscatter immersed (the points marked as BK and TX indicate the location of the backscatter and the transmitter, respectively)*

To emulate the functionality of an implanted sensor, a temperature measurement, taken with an external sensor, is transmitted using the LoRa packet OOK modulation described in section 3.2.2 of Chapter 3. Figure 5.21 shows the received data obtained with packet transmission over 60 min. The transmitter is located 80 cm away from the vessel with the backscatter immersed in the phantom liquid. The receiver is located at a distance of 1.5 m. In this test, the SF configuration exhibiting the best range is set up (SF = 12), keeping the bandwidth (BW) to 125 kHz to not drastically increase the time-on-air of LoRa packets. To test the backscatter, the straightforward protocol described in Figure 3.31 of Chapter 3 is used to send the temperature test value. The average number of frames lost is 4%. These results pave the way for long-range wireless implanted devices based on LoRa backscattering communication in indoor environments.

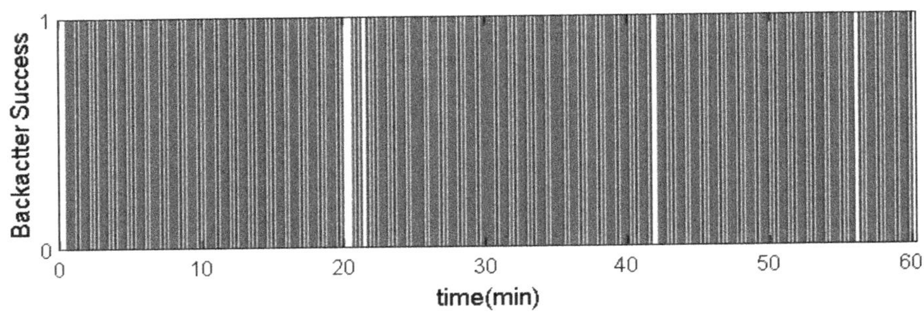

Figure 5.21. *Backscattered frames successfully received as a function of the time with the backscatter immersed in the center of the vessel with a phantom liquid and the transmitter located 0.8* m *from the vessel*

5.4. Discussion

The use of the 406–433 MHz frequency bands allows us to increase the coverage because the propagation losses are lower than at 868/915 MHz or 2.45 GHz ISM band (free space attenuation decays with the square of the frequency) and besides, the walls (Ali-Rantala et al. 2003) and indoor propagation (Tuset-Peiró et al. 2014) introduce lower attenuation at these frequency bands. This helps to counter the fading effect caused by multipath propagation in indoor environments which, in backscattering systems, affects the up and downlinks (Lorenzo et al. 2016, 2017; Milici et al. 2017a). The proposed system could easily use receiver diversity with multiple receivers to sniff shifted channels and prevent selective multipath fading. Channel hopping is not expected to provide significant improvements in the 406 MICS and 433 MHz ISM bands due to the small spacing between channels (less than the channel coherence bandwidth) so the channels are correlated (Lazaro et al. 2009; Tuset-Peiró et al. 2014). However, to prevent interference or collision with another system, a listen-before-talk (LBT) procedure can easily be implemented in the LoRa transmitter so that the available channels can be listened before the packets are sent. The parameters of the channel (frequency, SF, BW) can be transmitted to the receivers using a conventional LoRa link. This is different from the 2.4 GHz band, where channel hopping can be used to counter the effects of multipath propagation (e.g. in Bluetooth) (Tuset-Peiró et al. 2014).

Ref.	Technology /Power	Carrier Frequency	Modulation	Implanted Depth	Link Range Outside the Body	Bitrate (bps)	Base Station Receiver Sensitivity (dBm)
RamRakhyani et al. (2010)	Inductive /RF	700 kHz	CW (WPT)	20 mm	BS[6]	-	-
Lin et al. (2015)	Inductive /RF	13.56 MHz	LSK, ASK	15 mm	BS[6]	2 M	NA
Wentz et al. (2011)	Inductive /RF	10 MHz	PPSK, OOK	8 mm	BS[6]	1.35 M	NA
Lazaro et al. (2019)	NFC /RF	13.56 MHz	LSK, ASK	15–20 mm	3 cm	26.4 /848 k[7]	Depends on reader
Ma et al. (2018) Miozzi et al. (2019)	UHF RFID /RF	868/915 MHz		5-15 mm	0.6-1 m	40 k	-60/-70
Microsemi (2015)	Active /Battery	406/433 MHz 2.45 GHz Wakeup	OOK	-	10-20 cm	400 k	-91
Bae et al. (2011) Saadeh et al. (2017)	Endoscopy /Battery	20 to 925 MHz	BPSK, QPKS, FSK, OFDM	10–20 cm	BS[6]	-	-85
Khaleghi et al. (2018)	Backscattering /Battery	600 MHz	FSK	10 cm	BS[6]	1-5	-87
Besnoff and Reynolds (2012)	Backscattering /Battery	915 MHz	ASK	6 cm	BS[6]	30	-90
This work	LoRa Backscattering /Battery	406–433 MHz	LoRa	10-20 cm	<4 m	Few bps	-137

Table 5.4. *Comparison with other communication technologies for implantable medical devices*

In the following lines, there is a comparison of the technology presented here with other existing wireless technologies for implanted applications. Table 5.4 summarizes some examples of wireless technologies and compares parameters such as the carrier frequency, data rate, implant depth or read range. These technologies can be classified according to whether they are battery-less or not. Each technology has advantages and drawbacks depending on the requirements of the final application. Due to the higher losses of the body, wireless power transfer (WPT) based on inductive links at low frequency (LF) (RamRakhyani et al. 2010) and high frequency (HF) (Lin et al.

6 Receiver antenna on body surface.

7 The data rate depends on standard ISO15693 (26.4 kbps) ISO1443 (up to 848 kbps).

2015; Jiang et al. 2016) is the method preferred to power battery-less devices. Data are transferred by load modulation (Load Shift Keying – LSK) or by another wireless link at higher frequencies (Lazaro et al. 2019). Depth and power depend on the size of the coils used (Jiang et al. 2016), and the communication (read range) can be performed when the coils are close to the body. A drawback of this approach is that it requires specialized readers and ASICs. In Lazaro et al. (2019), the authors propose a method based on three coils to improve the depth and reading range using smartphones equipped with NFC and tags embedding commercial NFC ICs with energy-harvesting capabilities.

When the carrier frequency increases, the communication uses far-field and the attenuation increases very quickly with the frequency. Passive UHF RFID systems for identification and sensor communication at small implanted depths (5–15 mm subcutaneous) have been proposed (Ma et al. 2018; Miozzi et al. 2019). The communication is by backscattering, and a commercial UHF reader can be used. The WPT link is limited by the sensitivity of the UHF RFID IC (Miozzi et al. 2019) (about -20 dBm for modern ICs such as Impinj Monza R6 and -10 dBm for AMS-SL900A with sensing capabilities). The reader transmitter power is limited by exposure regulations (SAR) (Miozzi et al. 2019), so when body losses are higher, the read range is limited to 0.6–1.6 m depending of the depth of the implant (5–15 mm) (Ma et al. 2018).

Battery-implanted devices are used when high performance and high data rate transfer are required. Commercial active implantable transceivers are available on the market. For example, the Microsemi ZL70103 transceiver (Microsemi 2015) operates in the MICS band and the 433 MHz ISM band at a raw data rate of up to 400 $kbps$ and with wakeup at 3.45 GHz. Its sensitivity is -102 dBm at 200 $kbps$ (with a packet error rate of 10%), and the Tx/Rx average current is 5 mA. Several implanted wireless devices have been proposed for endoscopy (Bae et al. 2011; Gao et al. 2013; Liu et al. 2013b; Cho et al. 2015; Saadeh et al. 2017). For this application, to download video data, a high data rate is needed; therefore, active transceivers are used. However, the battery lifetime is limited to a few hours, which is enough for the exploration period required. Data rates from 1 $Mbps$ (Saadeh et al. 2017) to 80 $Mbps$ (Cho et al. 2015) have been achieved using different modulations such as BPSK (Cho et al. 2015), FSK (Bae et al. 2011), QPSK (Gao et al. 2013; Liu et al. 2013b) or OFDM (Saadeh et al. 2017), at frequency bands ranging from 20 MHz to 925 MHz transmitting power between 1 mW and 6 mW. The price to pay is high power consumption and, therefore, a limited lifetime. It is not the best option for long-term monitoring even if the data rate required is small.

Zero power backscatter gives better power consumption because it does not transmit RF power, which extends the life of the battery although the link attenuation is double that of active transceivers due to the forward and backscattered channels. The proposed backscatter can be used with other modulations such as ASK (OOK) or

FSK and the bit rate achieved (a few MHz) is high enough to transfer implanted video camera data. In this case, a CW tone illuminates the backscatter from the transmitter and the backscatter responds by modulating the carrier. A backscatter similar to that described in this study was used in Besnoff and Reynolds (2012) to achieve 30 $Mbps$ working at 915 MHz at a depth of 6 cm using a loop antenna. In another study (Khaleghi et al. 2018), a backscatter device implementing an FSK modulation achieved 1-5 $Mbps$ at 600 MHz and a depth of up to 10 cm. In both cases, the reader antennas were located close to the phantom, and the reader system was implemented using software-defined radio (SDR). The LoRa backscatter technique proposed in this study can achieve greater depths and read ranges outside the body because of the higher sensitivity of the LoRa modulation, but the main limitation is the low data rate. Another important point is that the proposed system does not require specialized readers because commercial low-cost LoRa transceivers can be used.

5.5. Conclusion

This chapter has shown the feasibility of communication based on receiving backscatter LoRa packets at a shifted channel for implanted devices. A proof-of-concept prototype has been developed. The prototype consists of an RF switch that presents two high reflective coefficients to an antenna controlled by a low-frequency modulating oscillator. A commercial ceramic antenna is tuned using a matching network at the MICS and ISM 433 MHz bands, which provides high miniaturization and high radiation efficiency. The backscatter can be interfaced with a low-power microprocessor to send low-rate data from sensors for monitoring purposes. Although the backscatter can be used with other modulations such as ASK and FSK, the proposed method uses the reflection of LoRa packets to exploit the high sensitivity and robustness of LoRa modulation; therefore, there is no need to integrate high power-consuming FPGAs or other ICs. A complete link budget has been presented which shows the potentiality of the communication technique. The method proposed achieves great implant depths that can be read at distances of up to several meters from the body, so that only one transmitter base station is enough to cover a typical room. Conventional LoRa transceivers are compatible with the proposed backscatter system, which can be used for long-range monitoring in indoor environments for future implanted and wearable devices.

6

Backscattering for IoT Applications

This chapter describes the development of an application created in the context of the Covid-19 pandemic, in which a smart mask was designed for automatic temperature monitoring and access control. This application is a clear example of the Internet of Things (IoT), where a conventional, non-electronic object is enhanced by adding connectivity and sensors to improve key aspects of the user's daily life.

6.1. Introduction

People's welfare became the major objective during the pandemic. With the arrival of the Covid-19 virus, measures to contain the spread of the disease among the population were taken. Confinement, use of face masks and gloves, temperature controls at the entrance of public places and social distancing were some of these measures, which stayed in place for a while. Authorities ordered the population to use wearable devices that complied with the quarantine measures, while telecommunication data was used to monitor crowds. Corporations worked on applications capable of confirming situations in which a person may have come into contact with newly diagnosed Covid-19 patients (Gupta et al. 2020). For this reason, smart mobile applications were proposed for monitoring purposes, such as monitoring employees in industrial environments (Kaiser et al. 2021).

However, the pandemic revealed the limitations of current technological deployments, particularly in what concerns the spread of the disease. IoT and smart connected technologies, in combination with data-driven applications, can play a critical role not only in the prevention, mitigation and continuous remote monitoring

For a color version of all the figures in this chapter, see www.iste.co.uk/lazaromarti/low-power.zip.

of patients, but also in the prompt enforcement of guidelines, rules and administrative orders to prevent future outbreaks (Gupta et al. 2020, 2021). The pandemic has accelerated the introduction of IoT in several fields such as E-health, smart hospitals, remote patient monitoring or supply chain management (Gupta et al. 2020, 2021).

The pandemic became a race against time, where the effectiveness of the aforementioned measures was continually verified. In order to prevent widespread contagion in public places, a protocol for measuring body temperature was established, isolating individuals exhibiting symptoms of Covid-19, such as fever. There are several methods for measuring body temperature (e.g. oral, rectal, axillary or tympanic). Some of these are compared in the literature (Ray 2017). Despite the number of options, most of them are not feasible outside of a medical environment and even less on a mass scale. Therefore, there is no question why the infrared thermometer was the method of choice for monitoring people's body temperature. An infrared thermometer allows a quick and non-contact measurement, but the mass use of this kind of thermometer created some latent drawbacks (Buoite Stella et al. 2020). These included the effect of the distance, the measurement site or the influence of the environment on the temperature readings. To ensure a correct determination of the temperature, the operators designated for the controls must follow a strict protocol, and anyone who undergoes the measurement requires an acclimatization time when coming from outside. In practice, the system is imprecise and inefficient, despite being the most suitable for mass population control.

Face masks allow the insertion of breathing sensors (Cheng et al. 2017), enabling individuals to carry out their usual activities despite the discomfort of wearing masks. The mandatory use of face masks to prevent the spread of coronavirus through the air has led to the adoption of masks equipped with sensors. Accordingly, a new generation of wearable devices, which receives the name of smart masks, has attracted the interest of various research groups (Ghatak et al. 2020; Kalavakonda et al. 2020; Kim et al. 2020; Pan et al. 2020; Fischer et al. 2021; Fois et al. 2021; Kalavakonda et al. 2021; Masna et al. 2021; Velluzzi et al. 2021). In Kim et al. (2020), the authors propose a personalized smart mask created from a 3D scanned face image, with the objective of minimizing skin irritation. Additionally, temperature and strain sensors have been integrated to detect irregular breathing, which is one of the symptoms of respiratory diseases. The strain sensor monitors possible facial irritation caused by the tight sealing of the masks. However, the temperature sensor is based on a thermistor connected to a microcontroller that exclusively measures the air temperature. In Kalavakonda et al. (2020, 2021), a smart mask senses particles of different sizes suspended in the air by using an onboard particulate matter (PM) sensor. Another example of a smart mask is proposed in Ghatak et al. (2020), where a self-powered device based on a textile triboelectric nanogenerator (TENG) is proposed to neutralize virus-loaded aerosols. Recently, Fischer et al. (2021) introduced a smart mask with wireless connection using Bluetooth low energy (BLE). It integrates a PhotoPlethysmoGraphy (PPG) pulse-oximeter and a pair of differential barometers, one internal and one external, which acts as a spirometer and are capable of measuring respiratory pressure during the breathing cycle.

The work presented in this chapter takes advantage of the acceptance of face masks and their widespread use during pandemics to incorporate new functionalities, introducing a novel approach to determine the temperature and breathing rate of individuals wearing smart masks. One of the key features of the proposed body temperature sensor is that it is not dependent on individual or ambient variations. Determining body temperature solely through skin temperature poses challenges due to the face's exposure to drafts and sudden changes in ambient temperature, which directly affect the measurements. In addition, since Covid-19-compatible symptoms can also manifest in breathing patterns, the mask also includes breath rate monitoring through an integrated airflow sensor. The smart mask data is transmitted via the LoRa backscattering technique presented in Chapter 3, ensuring low power consumption and consequently, a long battery lifespan. Furthermore, leveraging the work carried out in Chapter 4, a simple access and capacity control system is implemented. This system grants or denies access to public areas, such as academic, leisure, administration and healthcare centers, based on wireless readings from sensors embedded in the mask. Additionally, it can determine the subject's last position by analyzing the received signal strength. This approach eliminates errors introduced by operators and improves screening efficiency. The smart mask could also be used for monitoring patients while they wait in hospitals or medical facilities.

Sustainability must be an essential point to consider in any scientific advance. With an estimated monthly use of 129 billion face masks (Prata et al. 2020) during the hard periods of the Covid-19 pandemic, the design of a disposable electronic system integrated into the mask does not appear to be an environmentally friendly solution. The option of integrating the system into a reusable device has been taken into account. Although the electronics do not present disadvantages to be integrated into a dockable prototype, any modification of the mask could affect (or invalidate) the strict approvals that support it. For this reason, major changes in mask morphology have been avoided. Since this chapter aims to develop a proof of concept, the discussion of the effect of the system in reference to mask approvals is outside its scope.

The main contributions of the work described in this chapter are the design and integration of a dual heat flux sensor capable of determining the body temperature by using cheek surface temperature, the integration of an airflow sensor to monitor the breathing rate and detect coughing events, and the implementation of a LoRa backscattering wireless communication to extend the smart mask's lifetime, enabling, at the same time, the implementation of an automatic access control and monitoring system.

The chapter is organized as follows: section 6.2 begins with a brief overview of the design and operation of the system, and subsequently introduces the dual heat flux sensor used to measure the body temperature, the system for the breathing monitoring and the wireless communication based on LoRa backscattering. Experimental results are presented in section 6.3. This section starts with the procedure used to calibrate the dual heat flow sensor and continues with the results obtained from the measurements of temperature and breathing rate, as well as the detection of cough and positioning. Different non-contact temperature sensor technologies are discussed in section 6.4. Finally, the conclusion is presented in section 6.5.

6.2. System design

6.2.1. *System overview*

Figure 6.1 shows an overall system's operation sketch. The main goal is to filter people quickly and efficiently without the need for an operator. For this purpose, LoRa transceivers are proposed to be placed at the entrance of public venues. Low-cost LoRa radios ensure scalability and commercialization with a low budget. The transmitter is placed at the entrance, while receivers are placed inside and outside the site. This distribution of LoRa transceivers allows the implementation of a low-resolution localization system by means of the received signal strength (as described in Chapter 4). Using this method, it is possible to determine whether or not the subject has crossed the entrance and where it was last detected (exiting or entering). This property can be used to implement an automatic occupancy control system. The attenuation of the wall plays an important role in determining if the person is inside or outside. The use of two receivers has two main advantages, such as increasing the probability of receiving data packets under NLoS propagation conditions and determining if the subject has entered the building by comparing the RSSI of each receiver. The received power should be higher for the receiver closer to the tag, allowing it to track the smart mask as it moves. A simple classifier can be established from the RSSI or the SNR measured by the LoRa receiver for the received packets. Unlike other backscatter-based communication systems proposed in the literature, which require specific and often expensive SDR receivers, the system proposed here allows the use of commercial devices without having to make any modifications. Receivers work as a gateway, sending the data via WiFi to the cloud or a local server. In any system that handles personal data like this one, the privacy of the user must be respected. However, this should be managed at higher layers of the communication protocol by implementing encryption systems, which is currently beyond the scope of this work.

The smart mask integrates two temperature sensors to measure body temperature and monitor the subject's breathing. By placing the temperature sensor inside the mask, multiple measurements can be taken to check body temperature, unlike the single measurement provided by an infrared thermometer. This allows the system to obtain a set of temperatures and store them for further evaluation. The package payload consists of two main parts: the identifier and the sensor data (e.g. temperature or breathing rate). Alternatively, to reduce payload length, temperature and breathing rate can be replaced with an OK/NOK bit. The identifier is crucial for determining the origin of each mask's data, accepting valid identifiers and rejecting those that have crashed. To avoid packet loss due to collisions between multiple smart masks, anti-collision strategies have been implemented. The first strategy involves properly setting the LoRa parameters to reduce the communication range, capturing signals only from masks near the entrance. The second strategy randomly adjusts the waiting time between frame transmissions, minimizing potential collisions between frames from different backscatters. The system denies access to individuals with temperatures outside of the standard range, triggering an alert and preventing their entry. In case of irregular measurements, the mask identifier and temperature data are saved, providing a useful history and allowing the system to record nearby mask identifiers for tracking or possible infection alerts. Subsequently, a manual check of the patient's condition can be performed using conventional methods.

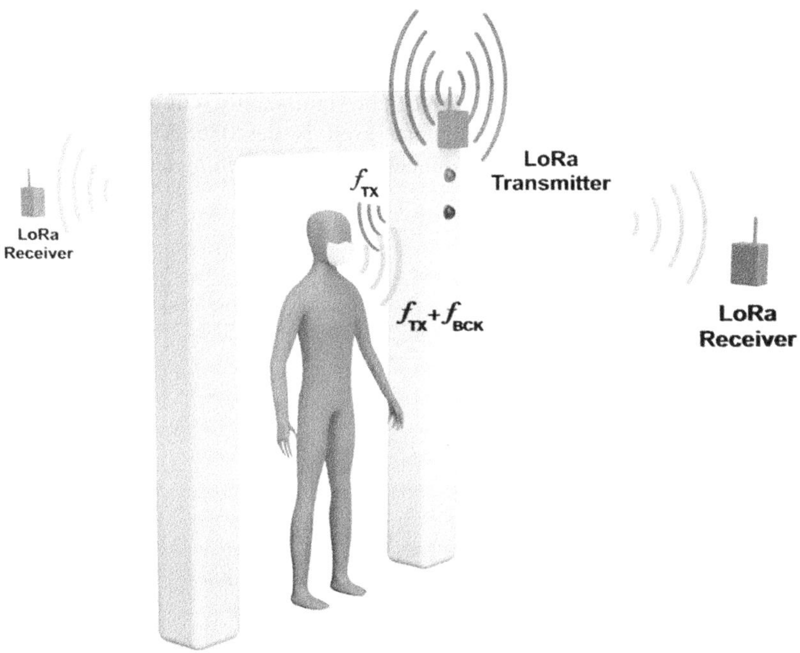

Figure 6.1. *Scheme of the system operation. LoRa transmitter is placed at the entrance illuminating the subject wearing the smart mask. LoRa receivers placed inside and outside receive the signal backscattered by the mask. Access is granted or denied depending on the value of the temperature readings*

6.2.2. *Implementation*

A schematic of the designed electronic device placed in the smart mask is presented in Figure 6.2. It integrates a dual-heat-flux (DHF) sensor for determining core body temperature and a thermistor to monitor the breathing cycle of the person wearing the smart mask. The low-power AtTiny402 microcontroller reads and processes the temperature data. Once the data are processed, it is sent to receivers via the LoRa backscatter uplink communication, as described in Chapter 3. Figure 6.3 shows an image of the prototype coupled to an FFP2 mask. The smart mask is powered by a 500 mAh lithium polymer (LiPo) battery. The system uses the load-modulation front-end, implemented with the previously mentioned ADG902. The combined power consumption of the microcontroller and the RF switch does not exceed 620 μA. The critical component in terms of power consumption is the DHF probe, which incorporates four MAX30205 ICs with a supply current of 600 μA each. Consequently, the system's maximum power can reach approximately 3 mA. In a worst-case scenario where all components are running at a 100% duty cycle, the battery life could extend up to 179 hours. The user can periodically recharge the battery once a week. Taking advantage of the MAX30205's shutdown mode with a current draw of less than 3.5 μA and the quick 50 ms temperature conversions, the overall power consumption of the system can be significantly reduced by duty-cycling the DHF sensor's operation. The antenna used is the 0868AT43A0020 ceramic antenna from Johanson Technology, connected to a Pi-matching network.

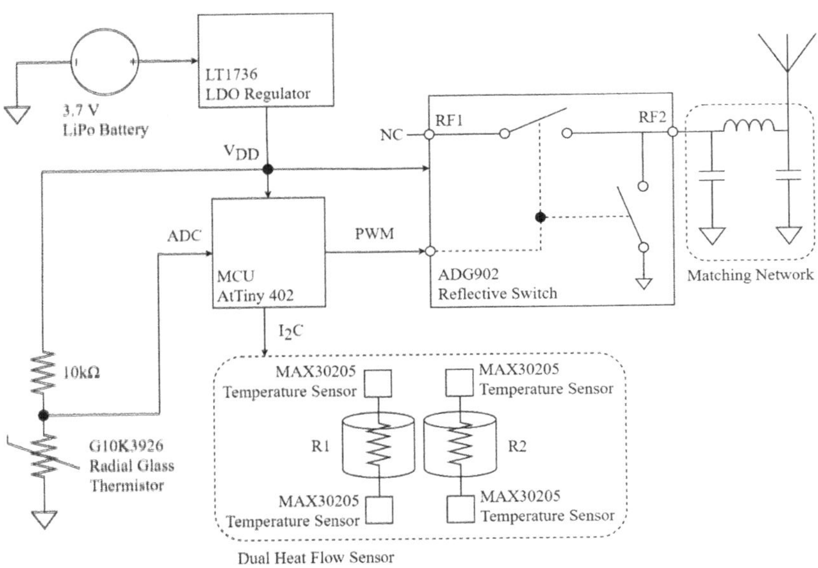

Figure 6.2. *Block diagram of the smart mask*

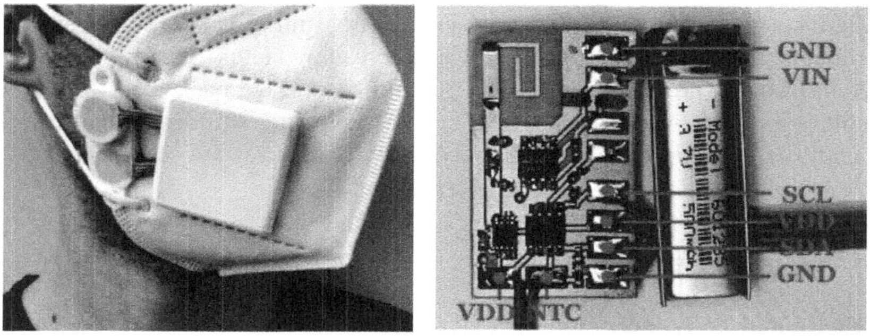

(a) (b)

Figure 6.3. *Photography of the smart mask (a) and the detail of the PCB (b)*

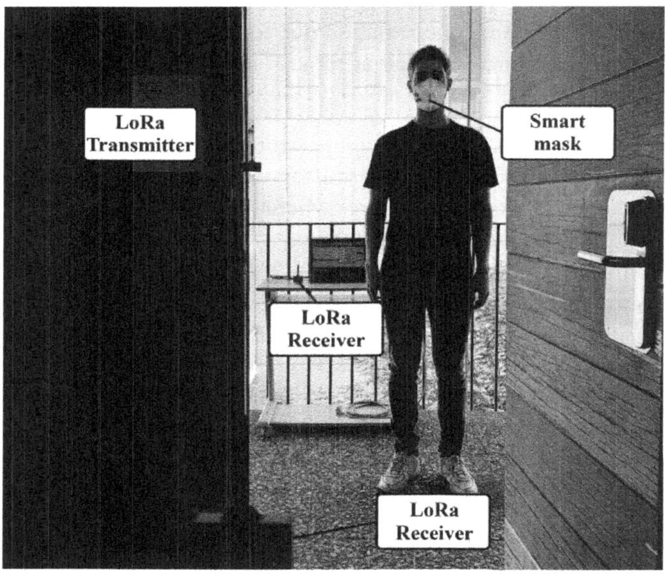

Figure 6.4. *Photography of the measurement scenario. LoRa transmitter is placed at the entrance. LoRa receivers are placed inside and outside of the laboratory*

Three low-cost LoRa32 boards from TTGO, operating at the 868 MHz ISM European band, are used for the transmitter and receivers located inside and outside of the laboratory, as shown in Figure 6.4. Both receivers are tuned at the sideband channel where the smart mask backscatters the LoRa packets. The transmitter is installed at the entrance, right in the middle point.

The distance between the receivers and the transmitter is set at 2 meters in the experiments. LoRa transceivers are set with a spreading factor (SF) of 7, a bandwidth of 250 kHz and a preamble length of 6. The air time of the packets is 18 ms. LoRa transmitter frequency is set to 868 MHz, and backscatter shifting frequency f_{osc} is set to 500 kHz. Consequently, LoRa receiver's frequencies are set to 868.5 MHz. The power of the transmitter is set to 17 dBm.

6.2.3. *Temperature sensing*

Determining body temperature through the skin has always been a challenge (Ilsley et al. 1983). Although conventional thermometers can perform this task accurately, the locations to measure the temperature must be very sheltered. The most common measurement places are the oral, tympanic, axillary and rectal. All of these are unfeasible for measuring temperature on a large scale or outside of a healthcare environment (Fulbrook 1997). Infrared thermometers have been the only method capable of handling the pandemic situation due to a quick and non-intrusive way of measuring body temperature, but their accuracy and reliability have been questioned in practice (Teran et al. 2012; Cui et al. 2020; Goh et al. 2021; Shajkofci 2021).

The simplest way to measure body temperature is by measuring skin temperature with a temperature sensor (e.g. with a thermistor or an active integrated circuit sensor). Although this method works well in places where the temperature is stable (e.g. in the armpit), it is not the most ideal system to be integrated into a mask, since it is highly exposed to drafts and changes in temperature. A method for determining deep body temperature is through heat flux sensors. Several works that analyze the performance of this kind of sensor can be found in the literature (Gunga et al. 2008; Kitamura et al. 2010; Feng et al. 2017). In Gunga et al. (2008), a single heat flux sensor based on temperature measurement at two points plus a correction to take into account the influence of ambient temperature is proposed. However, the calibration of the sensors depends on the thermal conductivity of the body. To solve this problem, DHF sensors can be used to measure the temperature in four points (Kitamura et al. 2010).

Two thermistors can be used to estimate the body temperature, one installed in the cheekbone and the other outside of the mask to measure ambient temperature (Lazaro et al. 2021b). However, with this methodology, it is necessary to calibrate the parameters of the model for each person because they depend on skin resistance. In addition, the thermistor must be calibrated with great precision. To avoid these problems, a DHF probe is designed and embedded into the face mask. Figure 6.5 shows the operating diagram of the core temperature probe. It consists of two heat flow sensors integrating four MAX30205 temperature sensors from Maxim Integrated, one MAX30205 pair for each heat flux sensor. This chip has an accuracy of 0.1°C and is specifically designed to measure the temperature of the human body. Both MAX30205 are isolated from each other with polylactic acid (PLA), which has a thermal conductivity equal to 0.13 $W/m \cdot K$.

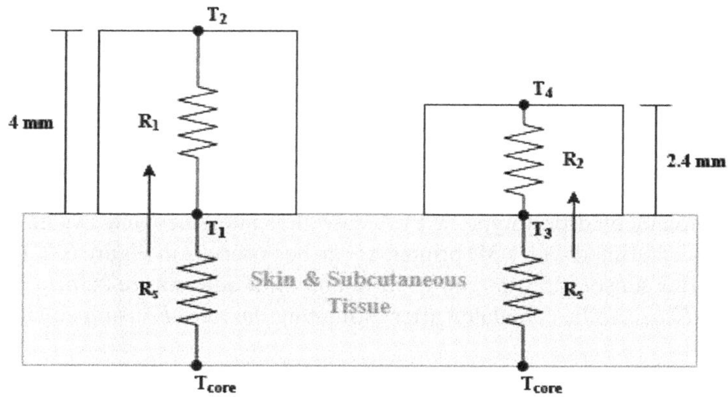

Figure 6.5. *Schematic diagram of the dual-heat-flux sensor and its equivalent circuit*

According to the second law of thermodynamics, heat will flow from the hottest point to the coolest point until both temperatures are equalized by diffusion. The DHF probe can be modeled with an electrical system (Kitamura et al. 2010; Feng et al. 2017), as described in Figure 6.5. In this analogy, temperatures are represented by the voltages in the nodes and the heat flow is the current that flows between two nodes. Analyzing this circuit, the following equations are obtained:

$$T_{core} = T_1 + \frac{(T_1 - T_2)R_s}{R_1} \qquad [6.1]$$

$$T_{core} = T_3 + \frac{(T_3 - T_4)R_s}{R_2} \qquad [6.2]$$

where T_1 and T_3 are the temperatures on the skin surface, T_2 and T_4 are the temperatures at the top surface of each heat flux sensor, R_s is the thermal resistance of the skin and the subcutaneous tissue and they have the same value for the two flux sensors. That is a reasonable approximation, considering that both sensors are very close to each other. The problem arises because the thermal resistance R_s of the tissue beneath each heat flux sensor cannot be measured and it is strongly influenced by the hypodermic blood flow and subject to variations between persons. The use of two heat flow sensors solves this problem. Eliminating R_s from equations [6.1] and [6.2], the core temperature can be obtained from the measured temperatures:

$$T_{core} = T_1 + \frac{(T_1 - T_2)(T_1 - T_3)}{K(T_3 - T_4) - (T_1 - T_3)} \qquad [6.3]$$

where K is defined as the ratio of thermal resistances R_1 and R_2 ($K = R_1/R_2$). The ratio K is determined during the calibration procedure from a known T_{core} and the measured temperatures:

$$K = \frac{(T_{core} - T_3)(T_1 - T_2)}{(T_{core} - T_1)(T_3 - T_4)} \qquad [6.4]$$

In the implemented prototype, a PLA case that integrates and isolates the four sensors has been printed with a 3D printer, as can be observed in Figure 6.6. The height of each heat flux sensor is 5 mm and 10 mm, and the diameters are 20 mm. The body temperature T_{core} can be calculated after estimating the ratio K using equation [6.3].

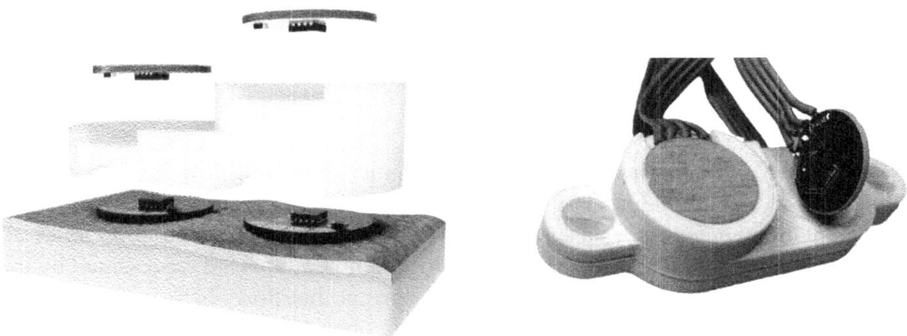

(a) (b)

Figure 6.6. *Dual heat flow sensor prototype built on a 3D printed PLA support, which acts as an insulator between the four temperature sensors: (a) 3D rendering of the prototype; (b) real prototype*

The work carried out for the design of the double heat flux sensor for body temperature determination has been leveraged to design a single flux sensor based on the same principle: two temperature sensors separated from each other by a known material. The designed heat flux sensor has been proposed as a cost-effective method to assess wall insulation, a topic of special importance to improve the energy efficiency of homes and move toward a more sustainable future (Lazaro et al. 2022c).

6.2.4. *Breathing monitoring*

Covid-19 disease can cause shortness of breath, lung damage and impaired respiratory function (Miller et al. 2020). Patients with Covid-19 can develop a complication known as acute respiratory distress syndrome (ARDS). With ARDS, patients lose the ability to breathe normally and this is known as a respiratory failure that results from severe inflammation of the lungs. Therefore, anomalies in breathing rate can be helpful for the screening of Covid-infected subjects. In addition, an increase in temperature is observed in people after having done some physical exercise (Xu et al. 2013). Therefore, tracking the breathing rate history can be useful in avoiding false positives.

Figure 6.7 shows the thermal behavior inside the mask (FFP2 type) during a respiration cycle. The temperature in the cheekbone and the lateral part of the mask remains mainly constant, whereas the central part of the mask changes according to the inhaled and exhaled air temperatures. Therefore, the DHF sensor must be installed on the lateral part of the mask to avoid heating associated with breathing. To monitor breathing, a temperature airflow sensor capable of detecting temperature variations is used. The temperature airflow sensor consists of a negative temperature coefficient (NTC) thermistor integrated into the mask and located under the nose. Compared to other temperature sensors, NTC devices are inexpensive and are characterized by both high sensitivity and high nonlinearity. In this design, absolute temperature is not required. Consequently, the calibration and the nonlinearity of this sensor are not critical issues. In the manufactured prototype, a radial glass NTC model G10K3976 from TE connectivity is used.

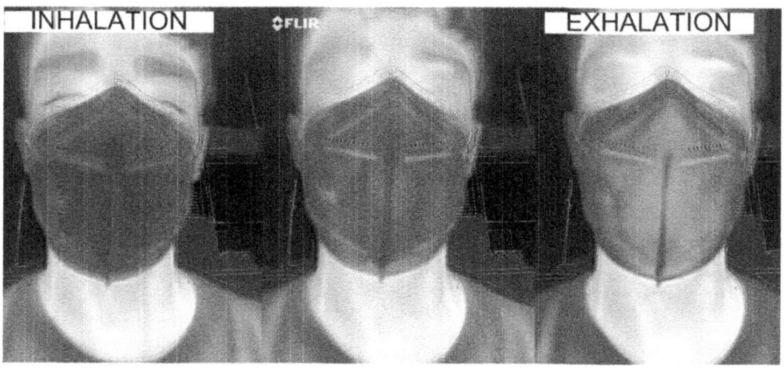

Figure 6.7. *Thermal image of a breathing cycle of a subject with a mask*

The first-order approximation of the Steinhart–Hart equation (Steinhart and Hart 1968) is given by the manufacturer to model the NTC resistance R as a function of the temperature:

$$\frac{1}{T} = \frac{1}{T_0} + \frac{1}{\beta} ln\left(\frac{R}{R_0}\right) \quad [6.5]$$

where T is the temperature in K, $R_0 = 10\ k\Omega$ is the nominal resistance at $T_0 = 298K$ and $\beta = 3976K$ is the Steinhart–Hart parameter, which is provided in the thermistor datasheet.

A simple voltage divider composed of the NTC and a resistance with the nominal value of the NTC (10 $k\Omega$) connected to the supply voltage V_{cc} is used to obtain the breathing measurement. The voltage output V_{out} of this circuit is acquired with the internal ADC of the microcontroller:

$$V_{out} = V_{cc}\frac{R}{R+R_0} \qquad [6.6]$$

In order to estimate the limitation in the temperature measurement due to the finite resolution of the ADC, the voltage sensitivity can be obtained by combining equations [6.5] and [6.6] and derive the result with respect to the temperature:

$$\frac{dV_{out}}{dT} = \frac{dV_{out}}{dR}\frac{dR}{dT} = -V_{cc}\frac{R \cdot R_0}{(R+R_0)^2}\frac{\beta}{T^2} \approx -\frac{V_{cc}}{4}\frac{\beta}{T^2} \qquad [6.7]$$

For temperatures in the order of ambient temperature $T \approx T_0$ gives an approximate value of -35 mV/K. Taking into account that the temperature variation due to airflow is on the order of 4–5 C, the microcontroller's 10-bit ADC provides enough resolution to digitize the output without the need for amplification, thus reducing the number of components.

This temperature signal estimated from the airflow sensor has a small variation over the average temperature of the mask; it is also noisy. Therefore, after the average signal is subtracted, low-pass filtering is applied to reduce noise. Since the filters must be implemented in a low-power microcontroller, exponential moving average (EMA) filters are used to reduce the computational charge and memory. The average temperature $X_{av}[n]$ is estimated using the following recursive equations from the measured temperature samples $X[n]$ (Milici et al. 2018):

$$X_{av}[n] = \alpha_1 X[n] + (1-\alpha_1)X_{av}[n-1] \qquad [6.8]$$

The output breathing signal $Y[n]$ is obtained from a second EMA filter applied to $X[n]$ after subtracting the average value to reduce noise:

$$Y[n] = \alpha_2(X[n] - X_{av}[n]) + (1-\alpha_2)Y[n-1] \qquad [6.9]$$

Typical values for α_1 and α_2 are 0.04 and 0.08, respectively. The breathing rate is usually expressed in breaths per minute (bpm). It can be estimated from the inverse of the interval between two consecutive peaks of the breathing signal. A robust peak detection algorithm described in Milici et al. (2017b) is applied to the filtered breathing signal $Y[n]$. When an interval between two consecutive breathings is longer than 10 seconds, it is considered apnea. In this case, an apnea index is activated to count the number of apneas during the measurement session. Compared to conventional airflow sensors used for breath monitoring, the integration of the airflow sensor in the mask allows for comfortable breath monitoring once the obligation to wear the mask is assumed.

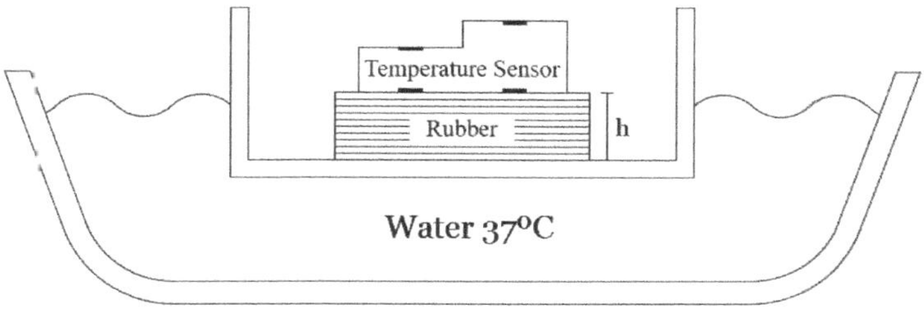

(a)

(b)

Figure 6.8. *Scheme of the body simulation method used to calibrate the DHF sensor (a), and photography of the experimental equipment used to calibrate the DHF sensor (b)*

6.3. Results

6.3.1. *Calibration of the temperature sensor*

In order to find the thermal resistance ratio K, an experimental setup similar to the one proposed in Nemoto and Togawa (1988) is conducted (see the schema in Figure 6.8). This experiment allows us to set the core temperature and check the calibration. A water bath at a temperature of $37°C$ simulates the nominal temperature of the body. A copper container is placed floating on the water. Rubber sheets are used to emulate the skin and subcutaneous tissue. The thickness of the rubber can be increased by adding more sheets. Rubber material has been chosen because its thermal conductivity $(0.17Wm^{-1}K^{-1})$ is close to the value of the subcutaneous tissue. Then, the probe is located on the rubber layer and the thermal resistance ratio K is determined for different thicknesses from the measurement of the four temperature sensors using equation [6.4]. The water temperature is monitored with a thermometer and controlled with a hot plate (with a resolution of $\pm 1°C$). A nearly constant value as a function of the thickness of the rubber layer has been obtained (see the top plot in Figure 6.9). The room temperature is $25°C$. Using the value of the ratio $K = 1.2$ that has been found, the water temperature is estimated from another set of measurements (see Figure 6.9). The differences produced could be attributed to poor lateral insulation of the prototype (Feng et al. 2017). Despite this drawback, the error is typically below $0.5°C$.

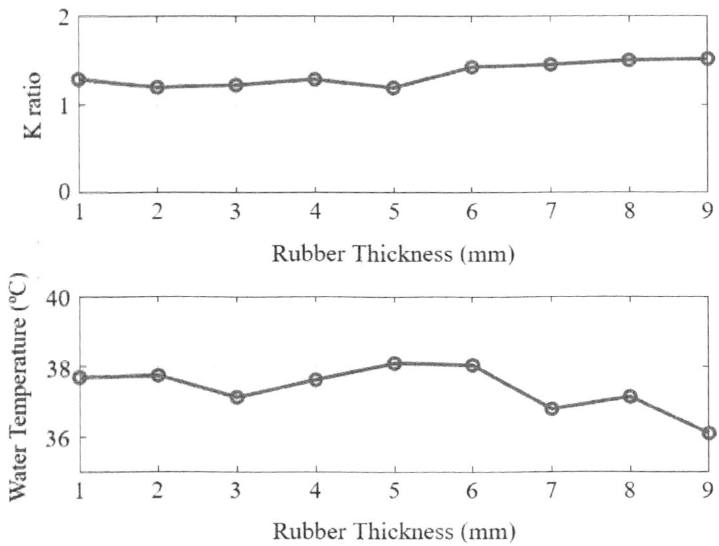

Figure 6.9. *Thermal resistance ratio K (top) and estimated water temperature (bottom) as a function of rubber thickness*

From the thermal resistance ratio K found with the latest experimental setup, the core temperature of a subject wearing the face mask can be estimated from equation [6.3]. A set of measurements are shown in the following figures. The experiments have been repeated with different persons, and similar conclusions have been achieved. The measured temperatures for each sensor of the DHF sensor, following the notation of Figure 6.5, are shown in Figure 6.10. The estimated core temperature and the measured temperature in the armpit with a thermistor are shown in Figure 6.11. It takes about 100 seconds to reach a value close to the core temperature. Figure 6.11 shows the difference between the two sensors once the temperatures reach the thermal equilibrium. The standard deviation error is $0.2°C$ which is of the order of the temperature sensor accuracy. When the mask is not worn, the four temperature sensors tend to ambient temperature and, therefore, it is useful to verify the correct placement of the face mask.

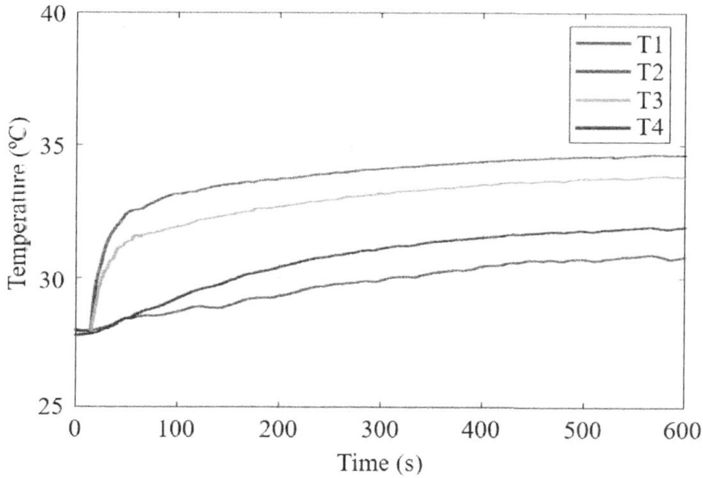

Figure 6.10. *Measured temperature for each temperature sensor in the probe as a function of the time after installing the smart mask*

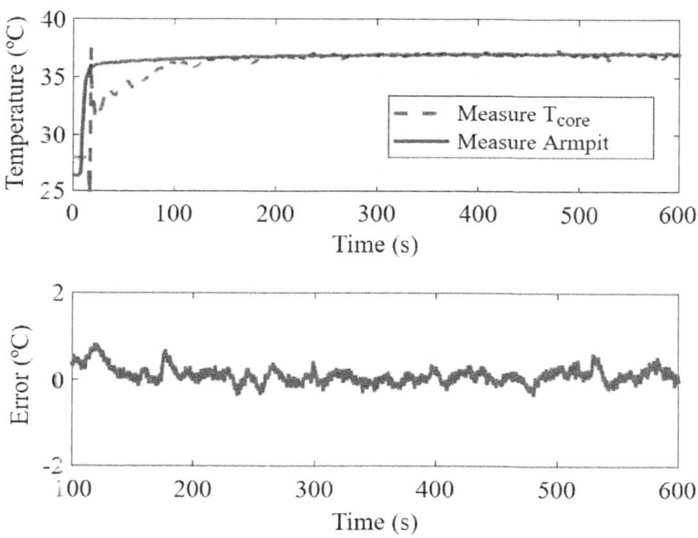

Figure 6.11. *Comparison of the estimated core temperature with the dual-heat-flux sensor and measured in the armpit with a thermistor (top) and difference between both measurements after the initial response (bottom)*

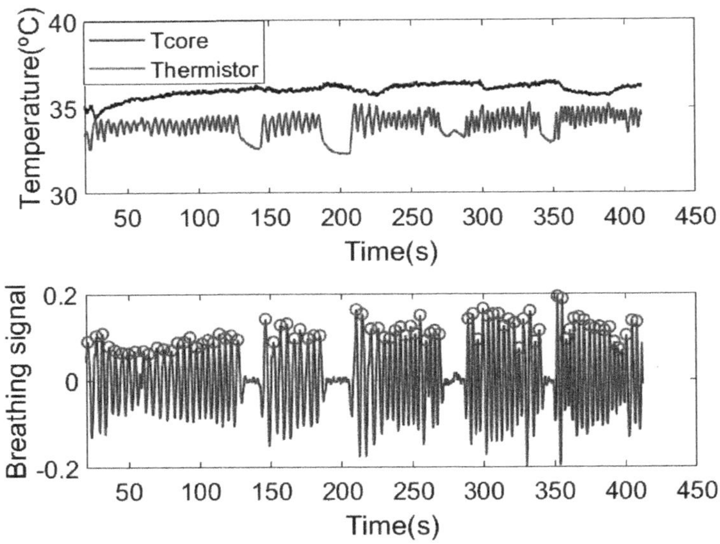

Figure 6.12. *Estimated core temperature and measured frontal thermistor temperature for breathing rate estimation (top). Breathing signal after removing and filtering the baseline (bottom)*

6.3.2. *Determination of breathing rate and coughing events*

An example of breathing rate estimation is shown in the following figures. Figure 6.12 shows the measured temperature by the frontal thermistor installed in the mask close to the nose area, in addition to the core temperature estimated with the DHF probe. Some apnea periods have been simulated by consciously stopping breathing. The maximum and minimum peaks obtained with the peak search algorithm are also plotted on the graph. The breath frequency can be obtained from the time position of the maximum peaks (see Figure 6.13). The average breathing is also shown in this figure. Apneas can be identified if the time interval between two consecutive breaths is greater than 10 seconds. An increase in the breathing amplitude is observed after the end of each apnea period. The breath frequency can be obtained from the time position of the maximum peaks (see Figure 6.13). The average breathing is also shown in this figure. Apneas can be identified if the time interval between two consecutive breaths is greater than 10 seconds. An increase in the breathing amplitude is observed after the end of each apnea period. These figures show how the smart mask can be used to monitor breathing activity. Some simple statistics such as the average breathing rate, the maximum breathing rate and the apnea count can be saved in the flash memory of the microcontroller. For continuous data logging, an SD memory card can be used to store the complete data over time.

The presence of coughing is another symptom associated with Covid-19. The system has the ability to detect cough events. Figure 6.14 shows the temperature of the thermistor after installing the mask. It can be shown that it takes about 25 seconds to reach a steady state. The figure shows the maximum peaks associated with the breaths. Under these conditions, the amplitude of the peaks remains nearly constant. To show the detection of cough, in this figure, the subject has coughed three times. Coughs cause a rapid rise in temperature of about $1°C$ due to the sudden expulsion of air. Therefore, by detecting these peaks and comparing the level with the average of the last five temperature peaks, a simple cough counter can be implemented (see the graph at the bottom of Figure 6.14).

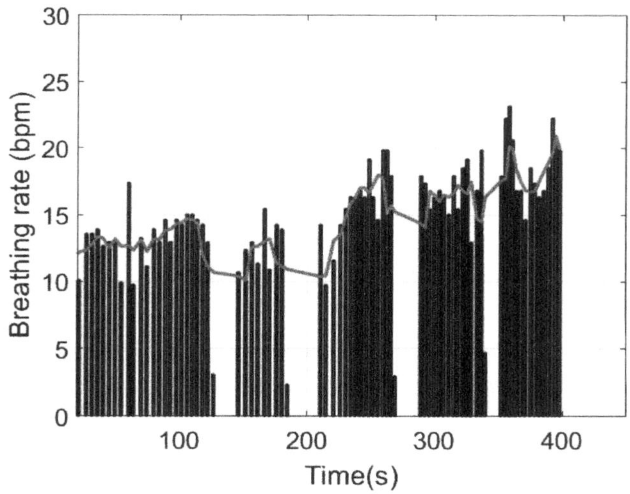

Figure 6.13. *Estimated breathing rate (vertical bars) and average breathing rate (red solid line)*

6.3.3. *Wireless measurement and positioning*

In Chapter 3, the LoRa backscattering localization system based on the received signal strength has been introduced. The work here proposed focuses on knowing the relative position of the person with respect to the door, establishing if the person is inside or outside the room. Therefore, a simpler classifier that does not require a training procedure is investigated. Some tests have also been done to evaluate the feasibility of locating the person from the backscattered signal using the setup shown in Figure 6.15, which corresponds to a typical access control (see Figure 6.4). LoRa ICs are equipped with measured received signal strength intensity (RSSI) and SNR indicators. The values of these measured parameters are stored in the corresponding registers and can be read via the SPI bus by the host processor. Figure 6.15 shows the decrease of the RSSI at receiver number one (placed outside) and the increase at receiver number two (placed inside) as the person walks through the door from outside to inside.

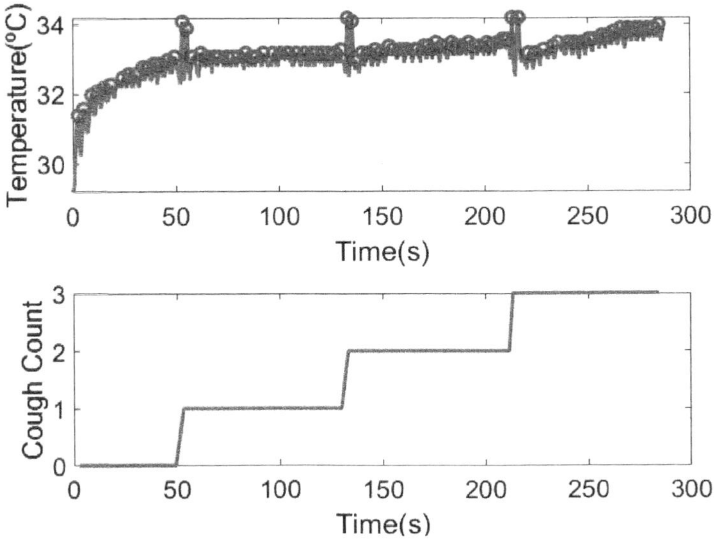

Figure 6.14. *Thermistor temperature measured after putting on the mask considering coughing episodes (top). Number of cough episodes detected from maximum temperature peaks (bottom)*

However, from equation [4.1] of Chapter 4, the received power would be expected to increase significantly as the person gets closer to the transmitter. Although there is a slight increase, it was expected to be larger. This is attributed to both diffraction effects and the blocking of the direct line of sight produced by the walls and the body itself. The measured RSSI for each receiver presents the crossover at approximately the equidistant point between each receiver and the transmitter. Following equation [4.1], the power at the receivers should be equal for equal distances between the transmitter and each receiver. This defines a plane located midway between both receivers. However, due to the attenuation of the walls, slight variations in the positioning of the receiver and, in general, a non-symmetrical scenario, a displacement of the crossover point may occur. A threshold difference parameter ($\Delta RSSI$) to establish the crossover point at the door entrance can be considered. The smart mask will be located inside when the difference between the RSSI measured at each receiver is greater than this parameter $\Delta RSSI$:

$$RSSI_2 - RSSI_1 > \Delta RSSI \qquad [6.10]$$

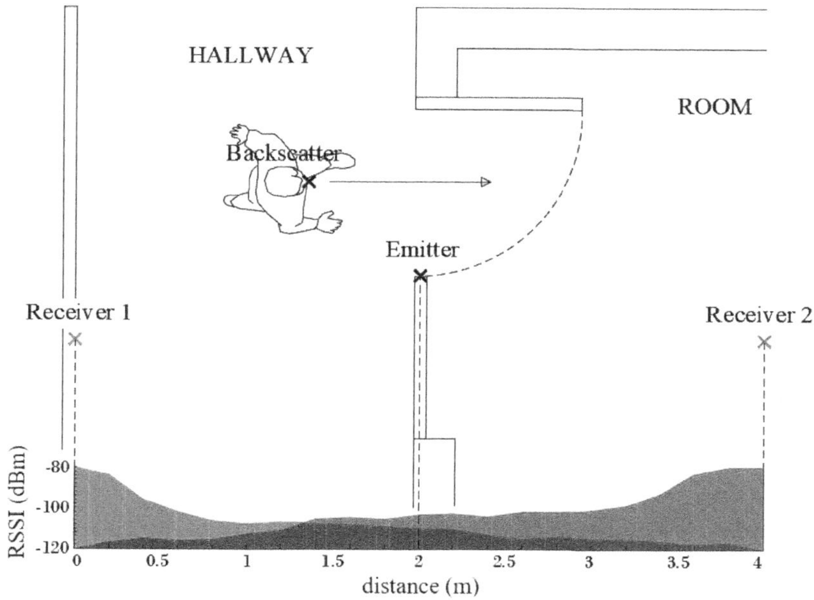

Figure 6.15. *Measured RSSI at the two receivers (receiver 1 outdoor and receiver 2 indoor) for a subject crossing a door, as a function of distance*

This parameter can be determined experimentally when the receivers are installed. The backscatter is placed in the center of the door, and then the difference of the measured RSSI values of the two receivers is averaged. An analog decision rule can be applied using the SNR measurements returned by the transceiver. To study the reception coverage, several measurements have been made with the backscatter device within a radius of three meters around the entrance. Figure 6.16 shows the cumulative distribution function (CDF) of the RSSI and the SNR received for points randomly taken both outside and inside the room. These figures show that the measured RSSI and SNR are higher for the receiver closest to the mask. The average SNR returned by the LoRa transceiver has a higher resolution than RSSI, so that, it is preferred to use the measured SNR values than RSSI values. Therefore, a simple classifier based on the measured SNR can be used to determine whether the smart mask is inside or outside the building or room. The smart mask is outside if the SNR of receiver 1 (outdoor) is higher than that of receiver 2 (indoor). The read range can be increased by varying the LoRa parameters and the power of the transceivers. The LoRa transceivers are connected to the Internet, and the information gathered is sent by message queuing telemetry transport (MQTT) protocol to a broker that can be accessed from any web browser or mobile app.

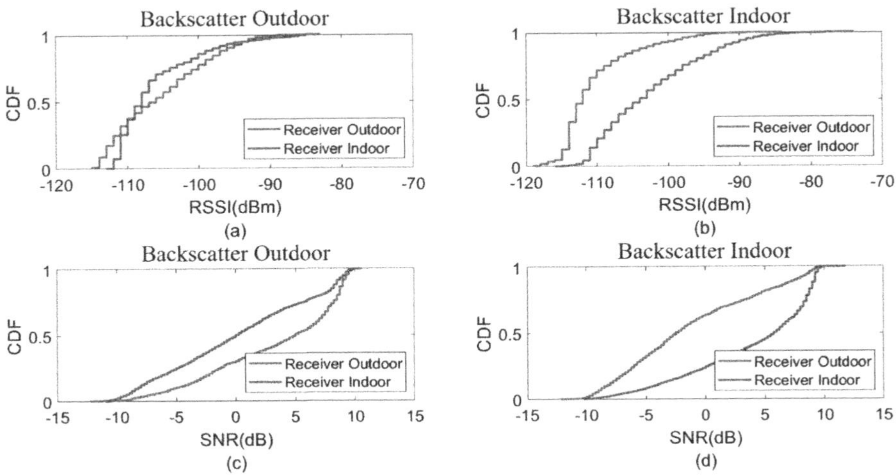

Figure 6.16. *Cumulative distribution function of the RSSI for (a) backscatter outdoor and (b) backscatter indoor, and SNR for (c) backscatter outdoor and (d) backscatter indoor*

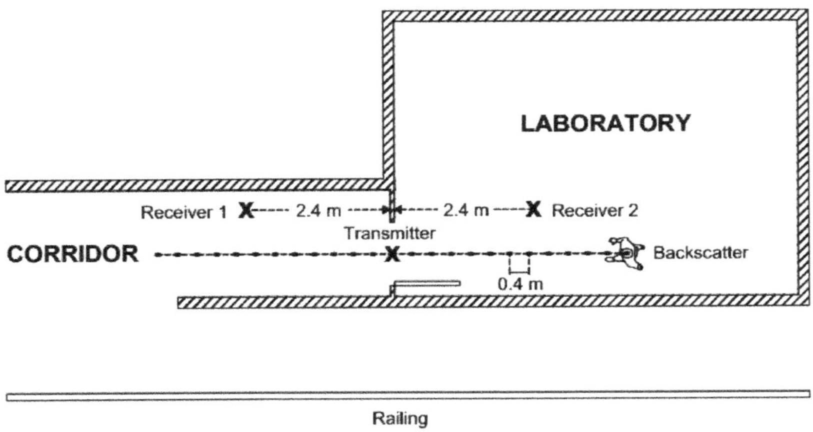

Figure 6.17. *Diagram of the scenario used in the measurements*

In order to investigate the performance of the positioning system, a number of measurements are performed in a scenario in which both line of sight and multipath propagation effects must be taken into account. Figure 6.17 shows a schematic of the scenario. It consists of a corridor that allows access to the laboratory through the door.

Each receiver is 2.4 meters away from the door and 1.5 meters above the ground. The transmitter is placed on top of the door. The entrance has two doors, one of which is made of metal and is kept closed. RSSI and SNR measurements are performed and recorded in each receiver in sets of backscattered packets measured every 40 cm along the corridor and the entrance of the laboratory, as shown in the diagram in Figure 6.17. A total of 1,000 measurements are done at each position with the smart mask worn by one person.

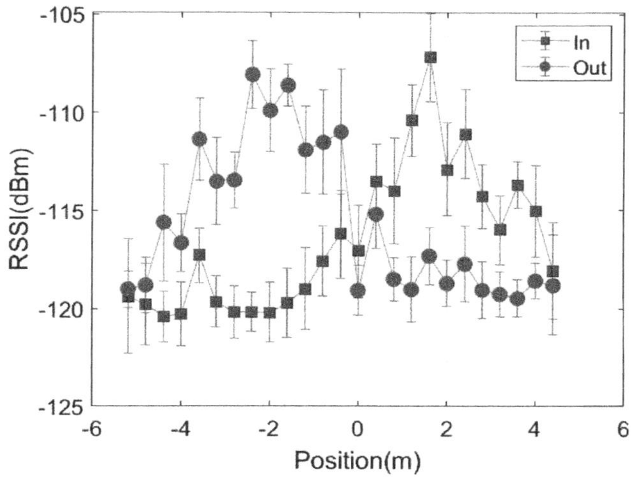

Figure 6.18. *Average RSSI measured as a function of the backscatter position with respect to each of the receivers. The standard deviation has been included in the error bars*

Figures 6.18 and 6.19 show the average RSSI and SNR values measured at each receiver (receiver 1 outside and receiver 2 inside), and the error bars show the standard deviation obtained. From these data, the decision rule is applied for each measurement cell. Figures 6.20 and 6.21 show the percentage of cases in which the RSSI or SNR measured at the receiver located in the laboratory is greater than that measured by the receiver located outside. Values close to 0% have been found corresponding to the positions furthest from the receptor. The result agrees with the measured RSSI and SNR values that reach their minimum levels, close to the sensitivity of the receiver. These cases can be ignored if they are considered to be outside the cover range. Again, slightly better accuracy is observed when using SNR measurements instead of RSSI measurements. Near the door, the accuracy is about 30% due to the measurement uncertainty and the random nature of propagation. Therefore, it can be estimated that the resolution is about one range cell (± 0.4 m), which is enough for this application. Therefore, the proposed decision rule to classify the relative position with respect to the door can be carried out without the need for a training procedure or the use of a specific propagation model.

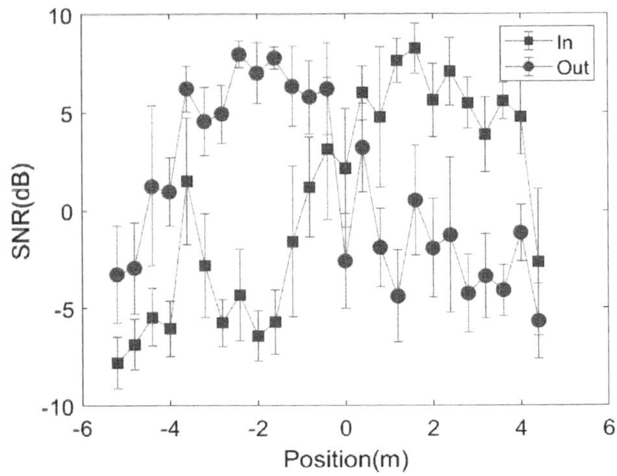

Figure 6.19. *Average SNR measured as a function of the backscatter position with respect to each of the receivers. The standard deviation has been included in the error bars*

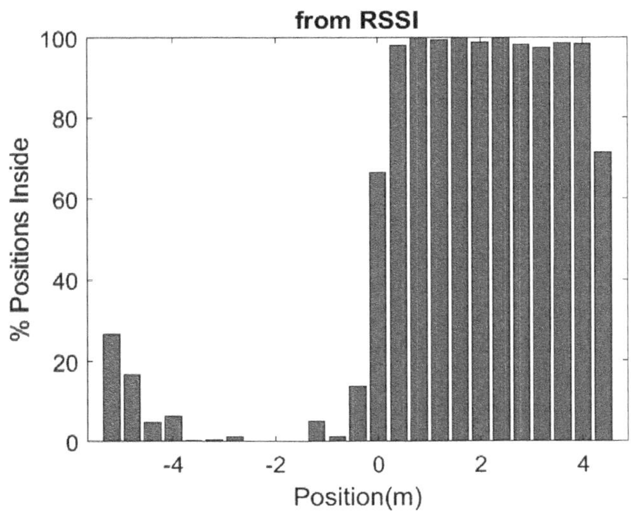

Figure 6.20. *Percentage of cases in which the smart mask is classified inside, as a function of the position with respect to each of the receivers, obtained by comparing the measured RSSI of each receiver at each position*

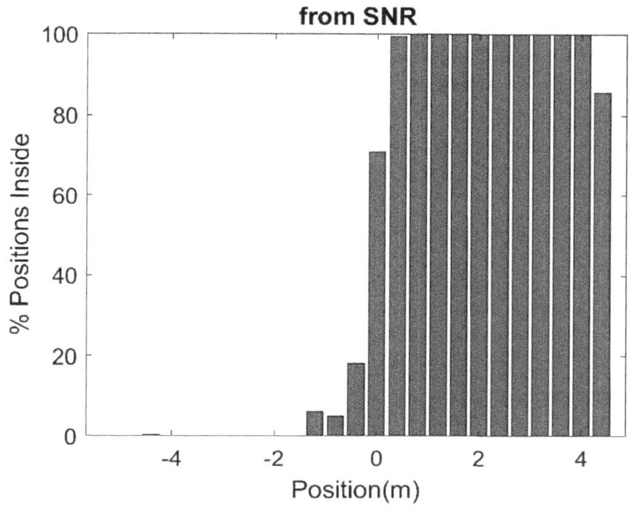

Figure 6.21. *Percentage of cases in which the smart mask is classified inside, as a function of the position with respect to each of the receivers, obtained by comparing the measured SNR of each receiver at each position*

To investigate the fluctuations of the RSSI (or SNR) depending on the position, a comprehensive empirical model has been developed. Propagation models applied to wireless signals within buildings have been extensively studied in different contexts such as cordless phones and wireless local area networks (WLANs) (Andrade and Hoefel 2010; Obeidat et al. 2018). The proposed model considers the effect of multipath propagation due to reflections on walls, floor and large objects, the shadowing or blockage of the transmitter and receiver due to the proximity of the body, and the attenuation due to walls and doors along the path. The model introduces an additional attenuation to the receiver power given by equation [4.1]:

$$L(dB) = L_{obs}(dB) + L_{multipath}(dB) + L_{blockage}(dB) \quad [6.11]$$

where L_{obs} considers the attenuation due to propagation through walls or other materials, and $L_{multipath}$ considers the attenuation due to the multipath interference, which is modeled using the two-slope model:

$$\begin{aligned} L_{multipath}(dB) =& 10 \cdot (n_T - 2) \cdot log\,(1 + d_T/R_{0T}) + \\ & 10 \cdot (n_R - 2) \cdot log\,(1 + d_R/R_{0R}) \end{aligned} \quad [6.12]$$

where n_T and n_R are the path loss exponents for distances greater than the breakpoint distance (Andrade and Hoefel 2010), and R_{0T} and R_{0R} are the backscatter to the transmitter and the backscatter to the receiver paths, respectively. This model does not consider any obstacle in the first Fresnel region, and the path loss exponent is close to 2 (or under 2 due to the corridor behavior, equivalent to a wave guide). On the other hand, a path loss exponent larger than 2 is expected when the distance is higher than the breakpoint. To account for the shadowing effect of the transmitter or receiver due to the proximity of the body, an increasing attenuation has been introduced as the backscatter approaches the transmitter or receiver:

$$\begin{aligned} L_{blockage}(dB) =& 10 \cdot n_{bT} \cdot log\left(1 + R_{1T}/d_T\right) + \\ & 10 \cdot n_{bR} \cdot log\left(1 + R_{1R}/d_R\right) \end{aligned} \qquad [6.13]$$

where n_{bT} and n_{bR} are the exponent coefficients that control the level of attenuation, and R_{1T} and R_{1R} determine the range extension for the blockage of the transmitter and the receiver, respectively.

Table 6.1 lists the empirical model parameters used in the simulations. Figure 6.22 shows a contour map of the received power (RSSI) simulated for each receiver for the same distances between transceivers and heights as in the scenario depicted in Figure 6.17. Figure 6.23 shows a cut for $y = 0$. Despite the limitations of the model, it justifies the evolution obtained in the measurements performed in this scenario. These figures show the ability of the model to predict the shadowing around the transmitter and receivers, and show the convenience of the decision rule [6.10] based on the highest RSSI (or SNR) that permits us to predict if the person has crossed the gate or not (position $x = 0$). Considering a deviation in the RSSI measurements of 2 dB, the resolution is on the order of ± 0.5 m, which is in agreement with the experimental results.

Parameter	**Symbol**	**Value**
Attenuation due to obstacles	L_{obs}	15 dB
Path loss exponent for the transmitter to the backscatter path	n_T	3
Path loss exponent for the backscatter to the receiver path	n_R	3
Breakpoint distance for the backscatter path	R_{0T}	1.5 m
Breakpoint distance for the backscatter to the receiver path	R_{0R}	1.5 m
Blockage exponent for the transmitter	n_{bT}	4.5
Blockage exponent for the receiver	n_{bT}	4.5
Extension of the blockage distance in the transmitter	R_{1T}	1.5 m
Extension of the blockage distance in the receiver	R_{1T}	1.5 m

Table 6.1. *Indoor propagation model parameters*

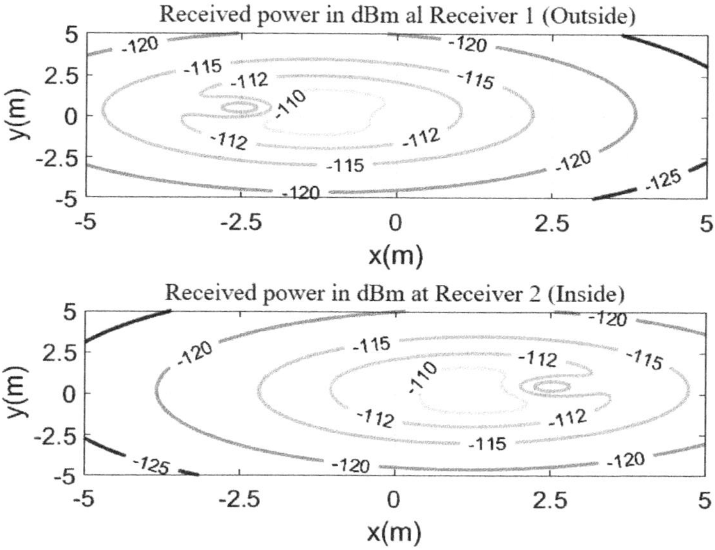

Figure 6.22. *Simulated received power (RSSI) at the receiver located inside (top) and outside (bottom)*

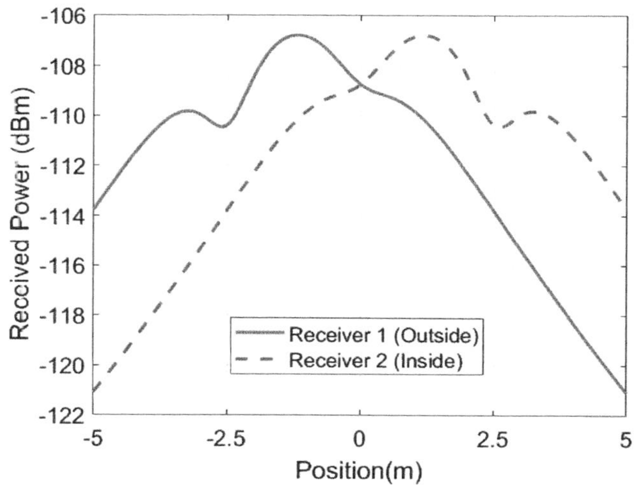

Figure 6.23. *Simulated received power (RSSI) as a function of the x-position at y=0*

6.4. Discussion

In this section, a comparison with other technologies (conventional and presented in the literature) is discussed. The comparison is summarized in Table 6.2. There are different commercially available methods for measuring body temperature. Body temperature measurement depends on the accuracy of the thermometer, the measurement site and the skill of the person measuring the temperature. The aim is to measure the core temperature, which is defined as the temperature of the large blood vessels in the internal organs and brain. Temperature measurement using a pulmonary artery catheter is considered the standard. However, this method is difficult to perform and is considered an invasive measurement method. Therefore, other methods and parts of the body such as the esophagus, the rectum, the armpits and the eardrum are used to measure body temperature (Lu and Dai 2009). Esophageal and rectal measurements are generally considered valid (Lefrant et al. 2003), but are not suitable for noncontact or continuous readings. Therefore, methods for measuring body temperature from measurements performed on its surface are considered noninvasive and are preferable for continuous measurements or home use.

Body temperature is measured on the surface of the skin and results from the ambient temperature and the temperature inside the body. The main problem is that it depends on factors such as blood circulation, activity and external temperature, and serves as a regulator of the body's temperature. Although the core body temperature remains mainly constant, the surface temperature of the human body can vary considerably (approximately 28–37°C). Depending on the ambient temperature, there can be big differences between the temperature of the core and the surface of the body. The temperatures in hands and extremities decrease several degrees in cold environments.

There are different ear or tympanic measurement devices commercially available. The most common are manual handguns, but recently there are also available portable over-the-ear devices for professional healthcare and for sports monitoring (e.g. from Cossinuss GmbH, Munich, Germany).

Axillary temperature measurement is a commonly used method. Modern sensors are based on a thermistor placed in the armpit, which is a place where the temperature remains stable and less influenced by the ambient temperature. Devices designed for continuous measurement are attached with the help of adhesive tape and can record the temperature and send the information via Bluetooth to a mobile application. Commercial examples of this technology are FeverSmart (model WT701, iMobile Healthcare LLC, Philadelphia, USA) and iThermonitor (model WT701, Railing Medical Company, Beijing, China). The main problem of these devices is given by the limitation of arm movements to prevent them from being disconnected or interfering with the measurement, as well as being problematic for measuring the temperature of babies.

Reference	Technology	Advantages	Disavantages
Commercial[1]	Infrared Thermometer	Cheap	- Operator needed. - Close detection. - Affected by external environment conditions. - Affected by body thermoregulation.
Commercial	Thermo-scanner	Remote detection	- Expensive. - Affected by external environment conditions. - Affected by body thermoregulation.
Commercial	Tympanic Thermometer	- Cheap. - Body core temperature estimation.	- Operator needed. - Close detection.
Commercial	Axillary Thermometer	- Cheap. - Body core temperature estimation.	- Close detection.
Yamakage and Namiki (2003)	Zero-flux method	- Body core temperature estimation.	- Commercially available - wired. - High power consumption for the heater.
Miozzi et al. (2017) Camera et al. (2020) Miozzi et al. (2020) Occhiuzzi et al. (2020)	UHF RFID	- Cheap epidermal tag.	- Expensive reader. - Position dependent. - Can be Affected by external environment.
Huang et al. (2016) Feng et al. (2017)	BLE and Dual Flow heat	- Wireless measurement. - Body Core temperature estimation - Reading from a smartphone	- Require a head band. - Does not measure breathing or location. - Requires authentication. - Long-range.
Jiang et al. (2019)	NFC	Cheap. Mobile as a reader.	- Measures the skin surface temperature. - Position dependent. - Affected by external environment conditions. - Affected by body thermoregulations. - Short-range.
This work	LoRa Backscatter	- Cheap. - Body core temperature estimation. - Breathing rate estimation. - Low resolution Localization. - Based on commercial LoRa transceivers - Integrated in a mask. - Long-range. - Cough detection.	- Battery assisted. - Low data-rate.

Table 6.2. *Body temperature measurement methods*

In addition to those places where the body surface temperature is similar to the core temperature (such as armpits or eardrums), there are other strategies based on estimating the core temperature obtained from heat flux measurements, such as the one considered in this application.

Zero-heat-flux (ZHF) method is a non-invasive method (Fox et al. 1973) that is commercially available, for example, 3M commercializes the model SpotOn (3M, St. Paul, MN). ZHF thermometers consist of a heater and two thermometers separated by a heat insulator. The heater is controlled to maintain the two temperatures identically. When these conditions are met, an isothermal tunnel from the core body to the skin surface is created. At this point, the subdermal temperature can be measured approximately 1–2 cm below the surface of the skin. In well-perfused parts of the body (such as the lateral forehead), the temperature of the tissue below the surface of the skin approaches the core body temperature. This non-invasive method is a useful tool for monitoring brain temperature. However, the current consumption associated with the heater makes this solution unsuitable for battery-powered or wearable devices. In addition, the patient must remain still to avoid the disconnection of the probes on the forehead.

As described in this chapter, DHF sensors are used as an alternative to ZHF sensors with similar accuracy (Kitamura et al. 2010). These sensors are also installed in the forehead and are commercially available for clinical use (e.g. Tcore product from Dräger, Lübeck, Germany). For instance, in Huang et al. (2016) and Feng et al. (2017), a DHF sensor has been integrated into a headband that allows data to be transmitted to a mobile phone via Bluetooth.

Infrared thermometers are a cheaper solution than infrared camera-based thermal scanners and are widely used to measure forehead temperature to examine people. A comparison of the accuracy of infrared thermometers with a tympanic sensor has been presented in Chen et al. (2020); it shows that there are fixed offsets that depend on the site where the measurement is made (forehead or wrist) and therefore, a suspicious person with fever requires a double check. In addition to the accuracy, the other issue arises from the need for a person to perform the measurement, and consequently, the system is not fully automatic.

In the last years, some research groups have developed epidermal temperature sensors based on UHF RFID tags (Miozzi et al. 2017, 2020; Camera et al. 2020; Occhiuzzi et al. 2020). The accuracy depends on the RFID integrated circuit, which embeds the temperature sensor and the RFID communication. These ICs have poorer sensitivity compared to conventional RFID ICs without sensing capability due to the extra energy required to bias the analog-to-digital converters as well as the sensors. In addition, losses introduced by the body limit the read range of these epidermal tags to a few meters. A prototype for the measurement of axillary temperature using passive RFID has been proposed in Miozzi et al. (2017).

1 Com.: Commercially available.

In Occhiuzzi et al. (2020), a dual-sensor based on a dual UHF IC embedded in a UHF RFID tag has been proposed to improve the estimation of the body temperature. Although the cost of epidermal tags is low, the readers are expensive and justified if the entire blockchain over RFID is deployed for other applications (e.g. to track the real-time location of assets, employees or customers), limiting the application for small users.

Recently, a large investigation can be found about battery-less sensors based on energy harvesting near-field communications devices (Lazaro et al. 2018). These RFID tags obtain energy from the RF field to bias low-power microcontrollers and sensors. Contactless epidermal sensors based on battery-less NFC (Lazaro et al. 2018) integrated into textiles have been recently investigated in Jiang et al. (2019). This approach solves the requirement of a specific reader because a mobile with NFC can be used as a reader, but the read range is limited to a few centimeters and gives the measurement of the shell temperature instead of the core temperature. Therefore, the accuracy and the utility for screening depend on the site where the tag is attached. When the tag is in the armpit, it can be accurate but it is not very practical for automatic screening applications.

The system proposed in this chapter has the advantages of embedding the core temperature sensor in a face mask as well as being fully automatic. On the other hand, the estimation of the relative location can be measured using low-cost infrastructure based on commercial LoRa transceivers. Therefore, an expensive reader and the installation of additional wearable devices are not required. The communication based on backscattering extends the battery lifetime. The main drawback is the low data rate due to the mechanism based on backscattering of LoRa packed that avoids the use of custom readers (e.g. SDR receivers). Therefore, for screening applications, basic information can be sent to avoid the amount of data (e.g. few bits to encode the temperature level, a bit to indicate the correct mask installation, a bit to indicate the usage time of the mask and a few bits to encode the breathing rate). The cost of the mask electronics is on the order of 5€, and a reusable future commercial device can be designed, changing the contaminated mask after use.

6.5. Conclusion

In this chapter, an automatic screening system based on a smart mask was carried out. The system copes with the problems of the infrared temperature measurement technology used during the pandemic such as the mandatory use of an operator and the low accuracy of the temperature measurements due to non-compliance with appropriate measurement conditions (distance, ambient temperature, conditioning time, etc.). The proposed system automatically monitors the temperature and the breathing of the subjects carrying the mask, allowing or denying access to public venues. Moreover, the system implements a simple localization system based on the received signal strength to determine if individuals are entering or exiting the building.

A novel DHF sensor for determining body temperature has been designed, calibrated and tested. The core temperature sensor output has been compared to the measurement of the temperature in the armpit, obtaining a small difference of $0.2°C$ in accuracy, enough for the proposed autonomous screening system. Furthermore, an airflow sensor based on a simple thermistor has been implemented along with an algorithm to determine the breathing rate and coughing events, allowing the detection of respiratory pathologies associated with Covid-19. Temperature and breathing data can be saved in the non-volatile memory of the microcontroller for further assessment. Additionally, the work described in this chapter can be extrapolated to other applications, for example, in industrial environments, integrating sensors to measure harmful gases.

Wireless communication has been implemented using LoRa backscattering, thus reducing the power consumption of the smart mask compared to proposals based on other wireless technologies such as Bluetooth. The high sensitivity of LoRa receivers ensures a communication range of several meters. Moreover, a simple access control system based on the received signal strength has been proven. Despite the intrinsic low data rate of the LoRa packet OOK modulation, it has been proven to be enough to receive the identifier and data of the smart masks close to the entrance. However, further work should be done to increase the backscattering communication throughput.

7

Backscattering for Domotic Applications

This chapter describes the design of a home surveillance system based on LoRa backscattering. This approach reduces the power consumption of the nodes, making them battery-free and consequently eliminating the need for periodic user maintenance. The chapter delves into the proposed modulation scheme for this application and presents measurements in real-world environments, thereby validating the feasibility of the system.

7.1. Introduction

The rapid advancement of technology has transformed home automation from a luxury accessible only to a few into a widely available solution. Developments in the Internet of Things (IoT), reductions in manufacturing costs and the standardization of protocols have enabled virtually anyone, even those without prior knowledge, to automate their homes. With the advent of artificial intelligence, it is now rare to find a household without a voice assistant, such as Amazon Alexa, Google Assistant or Apple Siri. Automating home lighting is as simple as replacing old bulbs with smart ones and asking the voice assistant to synchronize them.

A subset of home automation focuses on home security. Unlike conventional security systems, which are typically installed and maintained by third-party companies, plug-and-play security systems have recently become commercially available. These systems are compatible with major voice assistants and common wireless technologies, gaining significant interest due to their ease of installation and the absence of monthly maintenance fees.

For a color version of all the figures in this chapter, see www.iste.co.uk/lazaromarti/low-power.zip.

Unlike voice assistants or smart bulbs, which typically have access to a power source, low-cost security systems often integrate two types of sensors: magnetic sensors for door/window monitoring and passive infrared (PIR) sensors for motion detection. Given the installation points (windows, doors, elevated positions, etc.), wiring the sensors can be tedious, making battery power preferable. This leads to the objective of this research: designing a low-cost security system based on LoRa backscattering that reduces the power consumption of wireless nodes sufficiently to eliminate the need for batteries and, consequently, the periodic maintenance required by commercial systems.

In addition to significantly reducing power consumption through backscattering techniques, we propose a novel, fully analog modulation system. This system leverages the frequency deviation detection functionality inherent to all LoRa transceivers to determine the origin of received packets. This approach enables the identification of a large number of tags using a single frequency channel and, consequently, a single receiver. Specifically, if each wireless sensor (tag) introduces a small, unique and predefined frequency deviation, the receiver can use this information as an identifier. This modulation technique ensures superior system responsiveness compared to any digital modulation, making it particularly advantageous for surveillance applications.

7.2. System architecture

Figure 7.1 provides an overview of the proposed system's operation. The setup includes a LoRa transmitter, a LoRa receiver and wireless nodes installed at key monitoring points, such as entrance doors and windows. The transmitter periodically sends LoRa packets in the 433 MHz ISM band. When a wireless node is activated (the magnetic sensor is triggered), it backscatters the incoming LoRa packets to adjacent channels (433 ± 300 kHz), where the receiver is tuned to detect the signal. The receiver can detect the transmission on either of the two sidebands. Thus, this system provides greater robustness against narrowband jamming attacks. Additionally, for wideband jammers, increasing the modulation frequency of the tag (Δf) requires the jamming device to increase its bandwidth by a factor of two ($2 \times \Delta f$) in order to interfere with the signal.

The proposed backscattering system uses a bistatic architecture, where the transmitter and receiver are positioned separately. Compared to monostatic configurations, bistatic systems offer several advantages, as described in Chapter 1. Path losses are generally lower, as they avoid round-trip propagation. Additionally, separating the transmitter and receiver reduces system complexity and minimizes interference between the two, addressing the "doubly near-far" problem (Choi and Kim 2015; Van Huynh et al. 2018; Hessar et al. 2019). This architecture also provides greater flexibility by allowing independent placement of the transmitter and receiver. Optimizing their locations – "or increasing the number of transceivers" – can help compensate for environment factors such as multipath interference or signal attenuation, significantly improving the system's range and reliability (Jia and Zhou 2021).

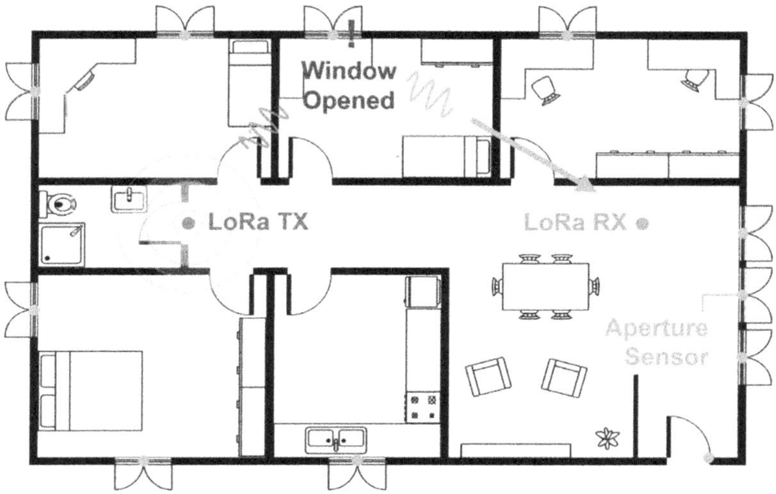

Figure 7.1. *Deployment diagram of the proposed system*

7.2.1. *Carrier modulation*

Home surveillance systems can operate across various frequency bands, ranging from 300 MHz to 2.4 GHz. Despite the existence of companies using private frequency bands, modulations and protocols, there is a prevailing trend towards standardization at 433 MHz due to its balanced trade-off between attenuation and throughput (Kumbhar 2017; Dong et al. 2016). Typically, commercial sensors use well-established modulation techniques such as amplitude shift keying (ASK) and on–off keying (OOK). These methods maintain a low transceiver complexity, resulting in reduced costs. However, it is important to note that amplitude modulation is susceptible to noise and interference, which may seem counterintuitive given its widespread use in wireless security systems. Countering jamming attacks is a difficult task. While any known wireless transmission can be interfered with, there are modulations and techniques, such as band hopping, reception confirmation and spread spectrum modulations, designed to make this task more challenging. The work described in this chapter proposes LoRa as the carrier modulation, bringing several advantages to the already-in-use ASK/OOK modulation. Figure 7.2 shows the spectrum of an ASK signal transmitted by a commercial sensor, overlapped with the CSS signal transmitted by a LoRa transceiver. Recall that LoRa technology stipulates various predefined configurations to balance the trade-off between speed and coverage. LoRa transceivers allow for the configuration of multiple parameters of the CSS modulation, such as bandwidth and spreading factors (SF). Additionally, parameters such as coding rate, sync word, preamble length and cyclic redundancy checksum (CRC) can also be configured. Based on these settings, the airtime of the LoRa transmission can be calculated. For this application, the CSS signal is configured to maintain an equilibrium between airtime and coverage, using a bandwidth of 125 kHz, an SF of 12, a 4/5 coding rate, a preamble of 8 chirps and no CRC. With these settings, the airtime of a LoRa packet is 663.55 ms.

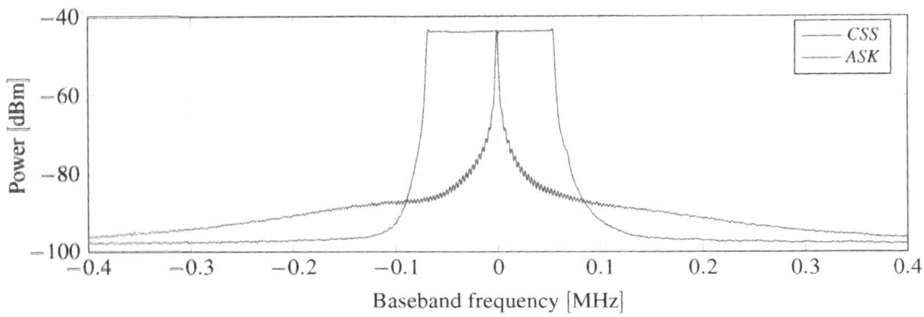

Figure 7.2. *Comparison of a commercial sensor transmission (Amplitude Shift Keying, ASK) with the proposed system's carrier modulation (LoRa) at 433 MHz*

7.2.2. *Tag identification*

Wireless security systems do not require a high data rate but necessitate a fast response, as they must alert the central unit when the sensor is triggered by sending a short packet. The goal in the development of this system is to significantly reduce the power consumption of the wireless sensors to make them battery-free. Based on this premise, the proposed system implements a completely analog modulation, thereby reducing the active time compared to a digital modulation and consequently lowering power consumption. For the implementation of this modulation, commercial LoRa transceivers, without any modifications, are used, making the system more cost-effective and efficient compared to applications that use an SDR as a transceiver.

Wireless nodes backscatter the incoming LoRa packets to the sideband when their sensor is triggered. To distinguish packets from each wireless node, each node can be configured to modulate at a different frequency channel (e.g. node 1: ±300 kHz, node 2: ±600 kHz, etc.), thereby avoiding packet overlap. However, this technique considerably limits the number of nodes that the system can manage, as it requires a receiver to listen on each channel. This, in turn, increases both the cost and the amount of spectrum used, making it impractical.

LoRa transceivers from Semtech (SX12xx, LR1121 and LLCC68) natively include a frequency correction function that compensates for oscillator drift. The receiver correlates the incoming signal with the previously known preamble of the packet (a train of up-chirps), synchronizing both signals for the demodulation process. The user can read the value of this deviation in hertz from the memory register. Leveraging this function, which is inherent in all LoRa receivers, all wireless nodes can backscatter packets on the same sideband channel, optimizing spectrum usage. Each node introduces an artificial frequency deviation that the receiver compensates for during the correlation process. By checking the frequency deviation, it can be determined which node sent the packet. Figure 7.3 shows the measured spectrum of the backscattered LoRa packets for five different nodes. The frequency offset introduced by each tag has been spaced to be visually perceptible; however, in practice, it can be set to be much lower.

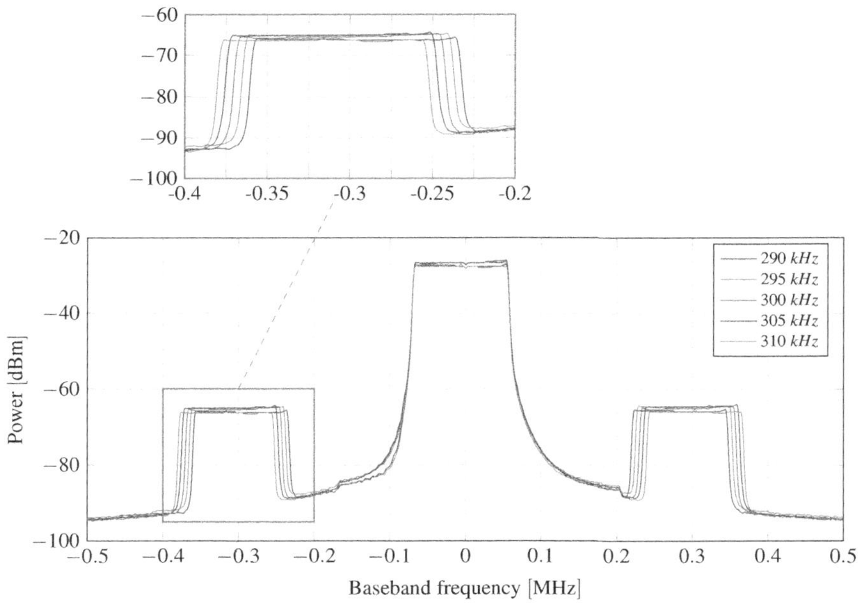

Figure 7.3. *Artificial frequency deviation introduced by each tag in the sideband channel. Carrier modulation: LoRa at 433 MHz; tag modulation frequencies: 290 kHz, 295 kHz, 300 kHz, 305 kHz and 310 kHz*

LoRa technology (CSS modulation), compared to other types of modulation such as frequency shift keying (FSK) or orthogonal frequency division multiplexing (OFDM), does not require precise oscillators, as the demodulation process can compensate for frequency deviation. For this reason, the oscillators in some LoRa transceivers exhibit considerable frequency drift. While this is not typically a problem under normal conditions, it poses an issue in this application, where we aim to identify wireless nodes by introducing a fixed frequency deviation. The Semtech datasheets indicate that LoRa transceivers can operate with a frequency deviation less than or equal to $\pm 25\%$ of the bandwidth (Semtech Corporation 2017). For instance, with a bandwidth of 62.5 kHz, the LoRa receiver would accept a frequency deviation of ± 15.625 kHz.

Many commercial LoRa modules integrate the SX127X transceivers, which exhibit a considerable drift in the oscillator frequency. Semtech released the SX126x transceivers as an upgraded version, incorporating a temperature-compensated oscillator (TCXO) that addresses the issues of its predecessor. Figure 7.4 shows the measured frequency deviation as a function of temperature for both transceivers. For this application, an SX1262 transceiver is used due to its oscillator frequency stability; however, the frequency of SX1276 transceivers can alternatively be compensated manually by performing a regression analysis of the frequency drift as a function of temperature. This approach requires integrating a temperature sensor with the transceiver.

Observing Figure 7.4, it can be seen that the frequency fluctuates by more than 977 Hz for the SX1276 transceiver (see Figure 7.4a), while it fluctuates by only 52 Hz for the SX1262 (see Figure 7.4b). As mentioned in section 7.2. on carrier modulation, LoRa transceivers support various configurations to balance the trade-off between communication range and throughput. For this application, a standard configuration is chosen, with a bandwidth (BW) of 125 kHz and an SF of 12. However, depending on the environment, the BW can be reduced to increase communication range, or the SF can be decreased to enhance throughput, and vice versa. With this configuration, the receiver supports a frequency deviation of the demodulated packet of ± 31.25 kHz. Considering the frequency deviation measurement of ± 26 Hz and adding an additional margin of 74 Hz, the system is capable of supporting a maximum of 625 sensors.

A comparison of the frequency deviation between the oscillators of the two main families of LoRa transceivers is conducted using two low-cost development boards. For the SX127x transceiver family, a TTGO LORA32 board is used, while a Heltec Wireless Stick V3 board is used for the SX1262 family. Both boards are configured for the 433 MHz band. Measurements are taken with the receiver placed indoors at room temperature, while the transmitter, equipped with a temperature sensor, is positioned outdoors for a duration of two days. This setup enables the measurement of fluctuations in the central frequency of the channel as a function of temperature.

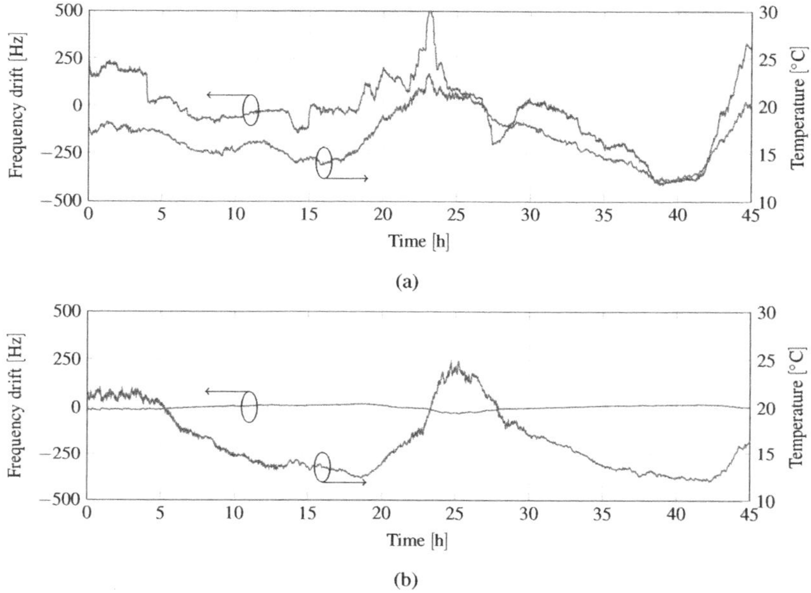

Figure 7.4. *Comparison of the measured oscillator frequency deviation (blue) as a function of temperature (red) for the SX127X family (a) and SX126X family (b) of transceivers*

7.2.3. *Tag implementation*

This section presents the design of the wireless backscattering node. The backscattering tag integrates the four parts described in Chapter 3, as shown in Figure 7.5: the reflection amplifier, phase modulator, control unit and energy harvesting/management unit. Each part is implemented separately for proof of concept, enabling the individual characterization of each part. The two blocks constituting the radio frequency front-end are fabricated on 32 mils Rogers 4003C substrate.

7.2.3.1. *Reflection amplifier*

The reflection amplifier is implemented using a BFR340F bipolar transistor from Infineon Technologies. In this case, a common-emitter topology with a negative feedback branch is used. The feedback branch is designed as a pi network, tuned as a bandpass filter at 433 MHz. Compared to the topology used in Chapter 3, this implementation highlights that multiple topologies can be considered for amplifier design. Figure 7.6a shows the schematic of the reflection amplifier.

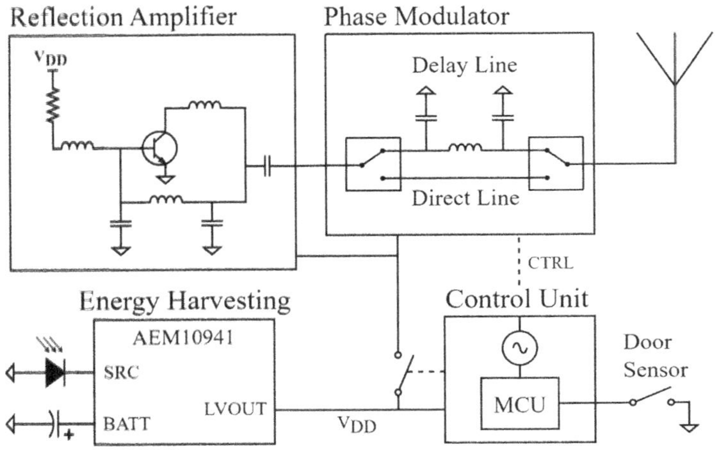

Figure 7.5. *Block diagram of the wireless node (tag)*

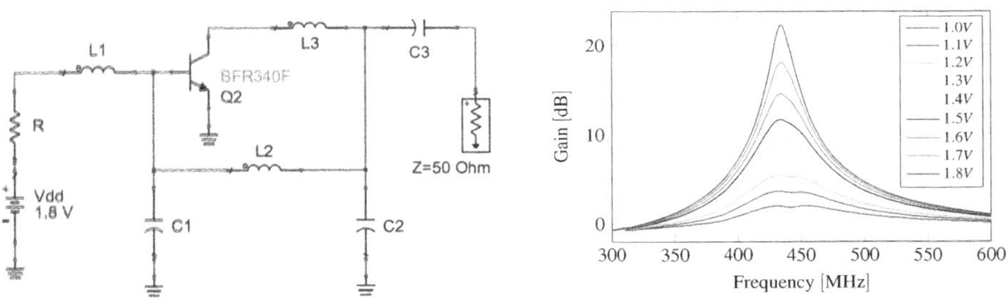

(a) (b)

Figure 7.6. *(a) Schematic of the reflection amplifier. Topology: common-emitter with a pi-network feedback loop.* $R = 300\ \Omega$, $L1 = 260\ nH$, $L2 = 15\ nH$, $L3 = 82\ nH$, $C1 = 15\ pF$, $C2 = 15\ pF$, $C3 = 33\ pF$; *(b) reflection amplifier gain as a function of bias voltage*

The reflection amplifier is characterized by the network analyzer configured with an output power of -50 dBm. Figure 7.6b presents its gain as a function of the transistor's biasing voltage. Figure 7.7a shows the amplifier gain as a function of the input power. It can be observed that it decreases as the input power increases, as expected due to the amplifier's gain compression. Figure 7.7b shows the current consumption of the amplifier as a function of the bias voltage. The current consumption and the gain of the reflection amplifier when polarized at 1.8 V are 322.9 μA and 22.3 dB, respectively.

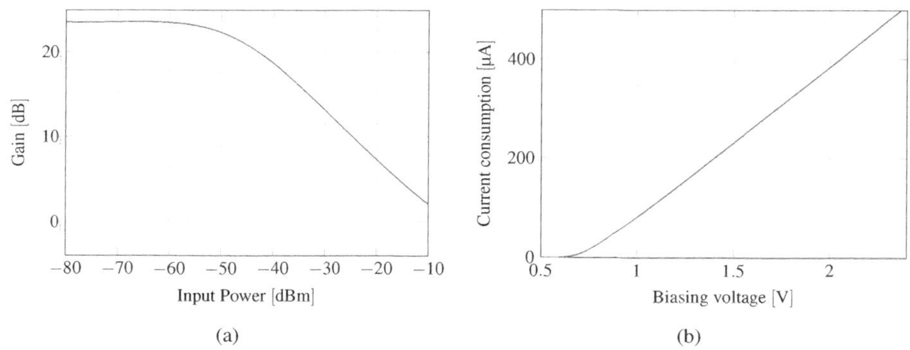

Figure 7.7. *(a) Reflection amplifier gain as a function of input power; (b) current consumption of the reflection amplifier as a function of bias voltage*

7.2.3.2. *Phase modulator*

Maximizing the differential radar cross-section involves generating two reflection coefficients that are 180 degrees out of phase with each other (Nikitin et al. 2007). To achieve this, a phase modulator is integrated between the antenna and the reflection amplifier, as described in Chapter 3. The phase modulator is implemented using two RF switches (ADG918), which switch between a direct line (no delay) and a delay line specifically designed to introduce a 90-degree phase shift at 433 MHz. As the signal passes through the modulator twice, it becomes 180 degrees out of phase compared to the direct line. The delay line can be constructed with discrete components or transmission lines. The first option introduces higher losses, especially at high frequencies, and is more complicated to implement due to the tolerance of discrete components. The second option introduces fewer losses and has lower manufacturing errors; however, at 433 MHz, a 90-degree phase shift line results in a significantly larger design. For this reason, discrete components have been selected. Figure 7.8 shows the schematic of the implemented phase modulator.

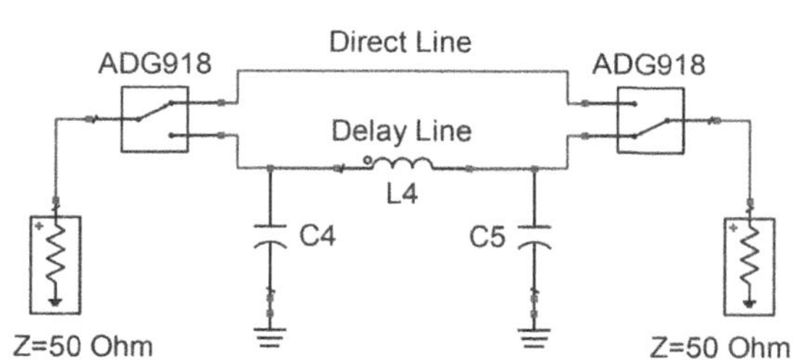

Figure 7.8. *Schematic of the phase modulator.*
$L4 = 18\ nH$, $C4 = 6.8\ pF$, $C5 = 6.8\ pF$

Figure 7.9 shows the characterization of the phase modulator. A phase shift of 84.28 degrees between the two states can be observed, quite close to 90 degrees, with an insertion loss of 2.43 dB and 3 dB for the direct line and delay line, respectively, at 433 MHz. It can be observed that the insertion loss for the delay line with the pi network increases significantly with frequency, making this topology not appropriate for designs at higher frequencies.

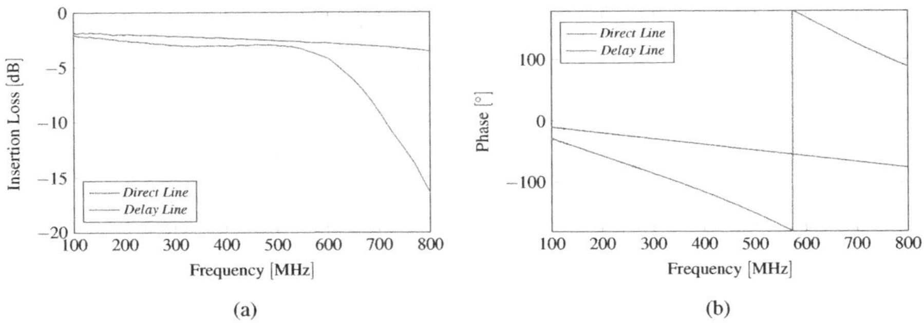

Figure 7.9. *Characterization of the phase modulator for both states: direct line and delay line. Insertion loss (a) and phase b) as a function of frequency*

7.2.3.3. *Energy harvesting*

There are several sources for energy harvesting (solar, thermal, kinetic, electromagnetic, etc.), but not all of them are suitable for any application. In the field of wireless sensor networks (WSN), the most common source is solar energy (Alsharif et al. 2019). Although this technique is not suitable for dark environments, such as garages, the continuous improvement in solar panel efficiency and integrated energy management chips has made it possible to harvest energy even in indoor environments without natural light. For the tag design, a light-based energy harvesting system is chosen, as this system does not limit the communication range of the tags. For proof of concept, an AEM10941 energy harvesting and management integrated circuit is used. This chip allows energy harvesting from an input power as low as 3 μW and a voltage of 380 mV. It also integrates a low-dropout (LDO) regulator with 1.2 V and 1.8 V outputs, making it ideal for powering low-power consumption ICs.

For this application, a 220 mF prismatic supercapacitor from CAP-XX (DMT3N4R2U224M3DTA0) is selected, as it provides a discharge rate very close to that of the lithium polymer battery while maintaining the longer lifetime of supercapacitors. The value of the supercapacitor is calculated based on several criteria. The minimum charge to which the supercapacitor's voltage can drop is the minimum required by the LDO regulator to maintain the 1.8 V output voltage. Thus, the supercapacitor voltage can range from 4.2 to 2.8 V (ΔV). The leakage current of the supercapacitor (3 μA), as well as the microcontroller's consumption in sleep mode (<1 μA), are taken into account when sizing the supercapacitor. Its value must be calculated within the most restrictive case of light exposition. This factor is not constant, as it varies depending on geographical coordinates. For example, in countries such as Italy or Spain, the shortest periods of daylight last about nine hours, whereas in northern regions such as Iceland, this can decrease to approximately four hours. Therefore, considering the most restrictive scenario, where the system must operate for 20 hours (Δt) without an energy source, the capacitance can be calculated mathematically by means of equation [7.1], assuming a discharge current ($I_{discharge}$) of 4 μA, with a resulting value of 206 mF. From the CAP-XX family of capacitors, the nominal value selected is the next available option above this calculation,

$$C = \frac{I_{discharge} \cdot \Delta t}{\Delta V} = \frac{(4 \cdot 10^{-6}\, A) \cdot (20\, h \cdot 3600\, s/h)}{(4.2\, V - 2.8\, V)} = 206\, mF \qquad [7.1]$$

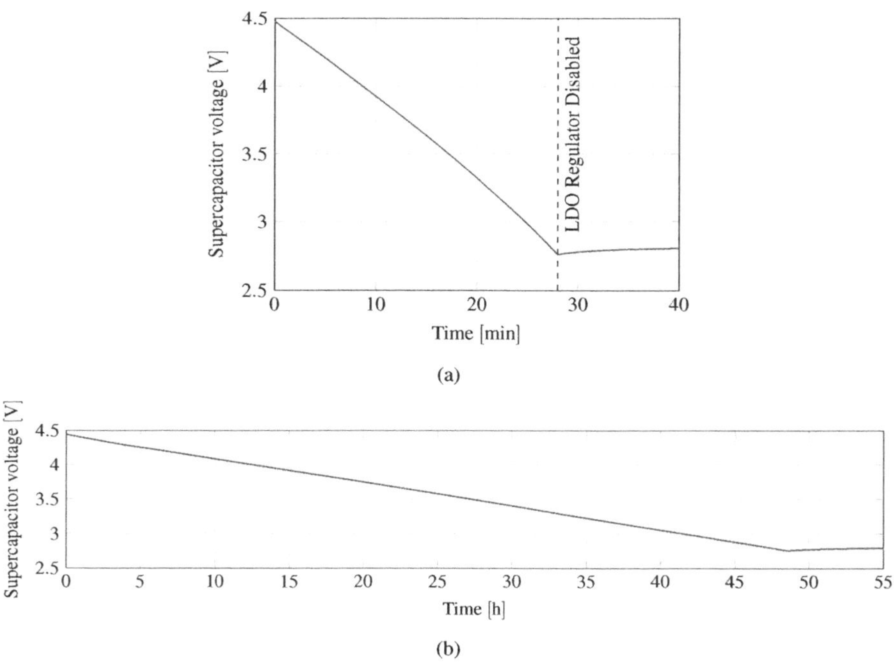

Figure 7.10. *Voltage of the supercapacitor as a function of time while the tag is in active mode (with the reflection amplifier enabled) (a) and in sleep mode (b)*

Figure 7.10a shows the voltage of the storage element (supercapacitor connected to the BATT pin) as a function of the operation time in active mode. When the sensor is triggered, the tag turns on the RF front-end to backscatter some LoRa packets towards the receiver before going back to sleep. Figure 7.10a validates the energy stored at the 220 mF supercapacitor, which is several orders higher than that required by the tag to trigger the alarm (less than a second in the best scenario). Figure 7.10b also shows the measurement of the supercapacitor voltage in the absence of an energy harvesting source, with the tag in sleep mode. The results validate that the system can remain powered for more than 20 hours without light.

It can be observed that when the voltage drops to 2.8 V, the LDO regulator output of the AEM10941 is disabled until sufficient energy is harvested to turn it on again. In the highest power consumption scenario (with the reflection amplifier activated), the tag can backscatter packets continuously for 28 minutes. In cases where the reflection amplifier is turned off (e.g. when the tag is close to the receiver or transmitter), it can remain in active mode for more than five hours. Considering that the time-on-air of a single LoRa packet is 663.53 ms for the settled configuration (SF = 12, BW = 125 kHz, preamble length = 8, coding rate (CR) = 1 and CRC disabled), when the sensor is triggered, it can send 2533 packets with the reflection amplifier active and 29,683 packets with the reflection amplifier disabled before running out of battery. In other words, accounting for a safety factor of 10, each time the sensor is triggered, the backscattering front-end will be active long enough to send 10 packets. Therefore, each sensor could be triggered 253 times with the reflection amplifier enabled and 2,968 times with it disabled before exhausting the supercapacitor charge.

7.2.3.4. *Control unit*

The control unit is responsible for activating the radio frequency front-end when the sensor is triggered. To achieve this, a low-power Attiny402 microcontroller from Microchip is used. The sensor connects to one of the I/O pins of the microcontroller, which remains in sleep mode most of the time and is activated by an interrupt when the sensor is triggered. Upon entering IDLE mode, the microcontroller powers the reflection amplifier, phase modulator and oscillator, which generates the tag's modulation signal. This signal can be generated in two ways: using the microcontroller's internal PWM or via an external oscillator. The first option is less costly, requiring only the microcontroller; however, it is less efficient in terms of power consumption, as generating a signal of 300 $\pm\Delta f$ kHz necessitates using the 1 MHz clock along with a prescaler, which increases the microcontroller's consumption. The second option requires an additional external oscillator but enables the microcontroller to operate with a 32 kHz clock. Despite the combined consumption of the microcontroller at 32 kHz (7 μA) and the external oscillator (26 μA), the overall power consumption is still lower than that of the microcontroller generating the signal with PWM (455 μA). Therefore, the second option is selected, using an LTC6906 oscillator from analog devices. Table 7.1 presents the current consumption of the polarized tag at 1.8 V, broken down by components. Table 7.2 details the cost of each component of the tag. The listed prices correspond to orders of at least 50 units; however, for large-scale production, costs would be lower. The total cost of the tag is estimated to be approximately 20 euros. Additional costs, such as PCB manufacturing, capacitors and inductors, may slightly increase or decrease the final price, depending on the supplier, the order quantity and the PCB surface finish. Figure 7.11 provides a rendered illustration of the four components described above, integrated with a 433 MHz SMD ceramic antenna.

Part	IC	Current [μA]	Power [μW]
Reflection amplifier	BFR340F	322.9	581.22
Phase modulator	ADG918	~2	~3.6
Oscillator	LTC6906	21	37.8
Microcontroller	AtTiny402	7	12.6

Table 7.1. *Breakdown of tag power consumption*

Section	Component	Cost/unit [€]
RF front-end	BFR340F	0.129
	RF Switch (ADG918)	2.93
	Oscillator (LTC6906)	2.38
	Ceramic antenna (0433AT62A0020E)	1.46
Energy harvesting	Energy managment IC (AEM10941)	2.55
	Monocrystalline solar cell	2.16
Control unit	Microcontroller (AtTiny402)	0.447

Table 7.2. *Breakdown of tag component costs*

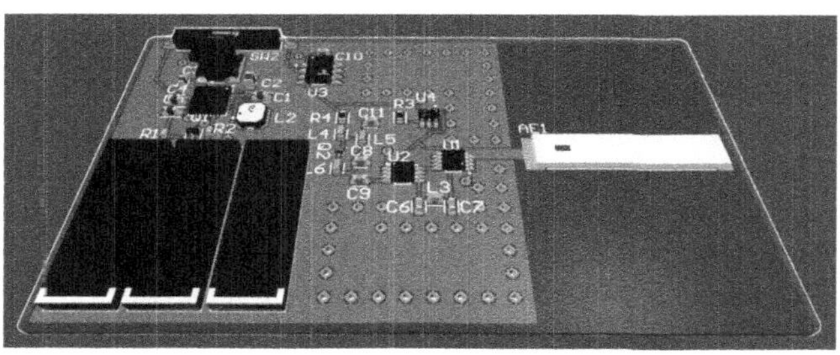

Figure 7.11. *Rendering of the printed circuit integrating the four parts described in this section: reflection amplifier, phase modulator, control unit and energy harvesting unit. The 3D rendering was generated using Altium Designer*

7.3. On-site measurements

Early experiments are carried out in the laboratory using monopole antennas at 433 MHz. Figure 7.12 shows a comparison of the spectrum when the reflection amplifier is turned on and off. Signals are recorded using an Adalm Pluto software-defined radio (SDR) as a receiver and a commercial LoRa transceiver as a transmitter. The transmitter and receiver are located at distances of 3 meters and 0.5 meters from the tag, respectively. LoRa is configured with an SF of 12 and a bandwidth of 125 kHz.

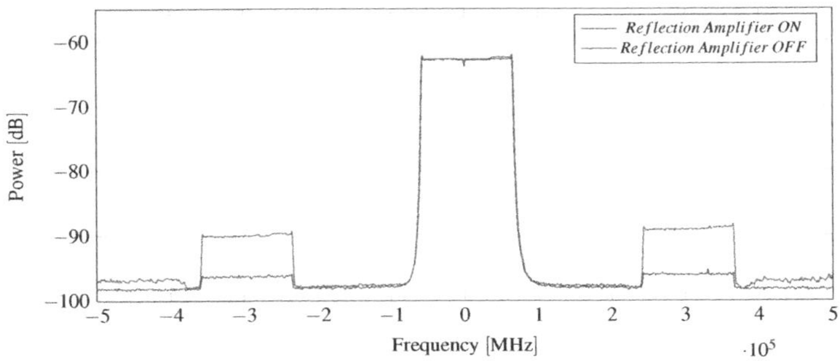

Figure 7.12. *Spectrum of backscattered sideband signals when reflection amplifier is turned on (red) and off (blue)*

To assess the system's resilience to interference, a broadband signal capable of interfering both backscattered sidebands is generated. A QPSK signal with a bandwidth of 1.3 MHz is transmitted with the signal generator, and its power level is progressively increased. The interference power is measured close to the receiver antenna using a spectrum analyzer connected to an antenna identical to that of the receiver to ensure consistency of the measurements. The spectral power of both the backscattered LoRa signal and the interference is evaluated within the 125 kHz-wide sideband channel. Figure 7.13 presents the spectrum of the backscattered LoRa signal along with the QPSK interference at different power levels. As anticipated, the receiver successfully demodulates LoRa packets even under negative carrier-to-interference (C/I) ratios, owing to the processing gain inherent to CSS modulations, which enables demodulation of signals below the noise floor. Figure 7.14 shows the packet error rate (PER) as a function of the C/I ratio. For C/I ratios equal to or greater than 0 dB, the PER remains below 3%. However, as the C/I ratio drops below 0 dB, the packet loss rate starts to increase rapidly. Specifically, for C/I ratios of -9.78 dB and -11.68 dB, the PER reaches 37.74% and 46.75% of lost packets, respectively.

To validate the accuracy of determining the provenance of a received packet (by checking the frequency deviation from the central frequency), a backscattering tag is modulated at 301 different frequencies. Each modulation represents one possible channel, with channels spaced 100 Hz apart. The channel separation of 100 Hz is set to be four times greater than the frequency error of the TCXO ($\pm 26Hz$), as determined in section 7.2.2 on tag identification. LoRa can demodulate packets with a frequency deviation of $\pm 31.25\ kHz$; however, the experiment is conducted within half of that range ($\pm 15\ kHz$) to avoid operating at the theoretical limit. The results indicate that the frequency deviation detected by LoRa commercial transceivers is influenced not only by the oscillator's frequency drift but also by the magnitude of the frequency deviation relative to the central channel. Figure 7.15 shows a diagram of the channels evaluated experimentally. In this application, each wireless sensor in the system is configured to transmit on a single, unique channel. Therefore, accurately identifying the channel is crucial for determining which tag is transmitting.

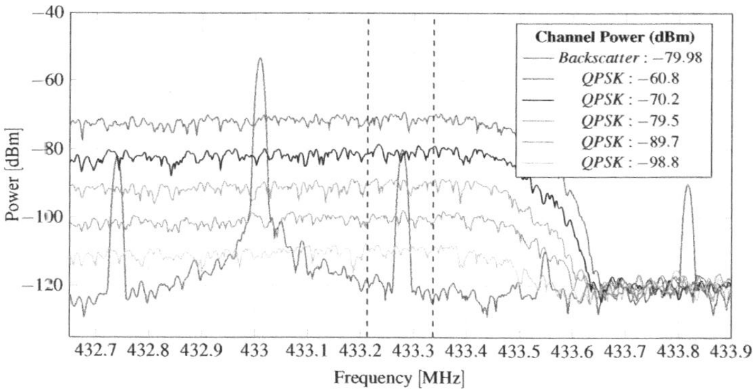

Figure 7.13. *Spectrum of backscattered sideband signal over different levels or interference (QPSK signal)*

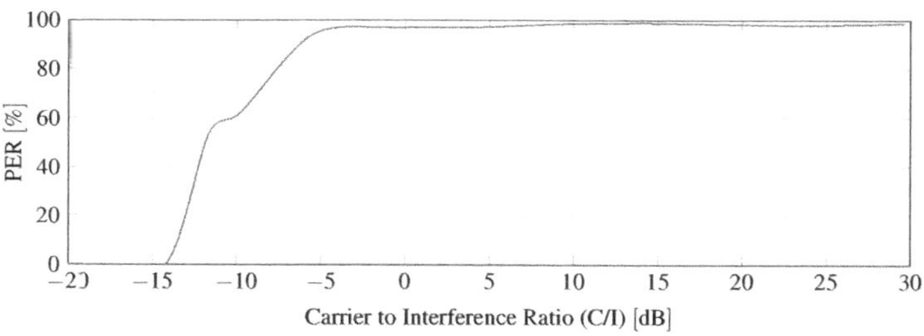

Figure 7.14. *Packet error rate (PER) as a function of carrier-to-interference (C/I) rate*

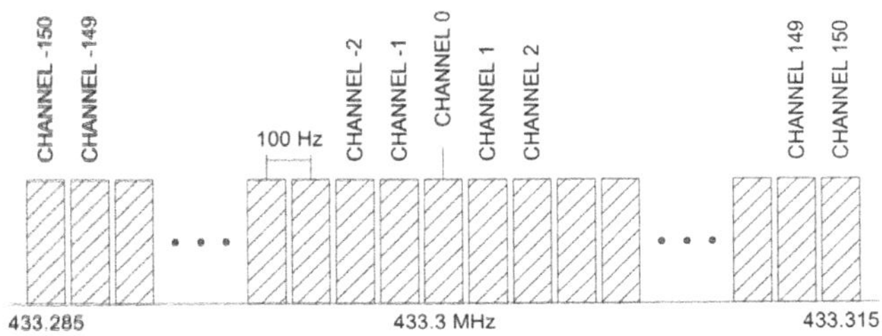

Figure 7.15. *Diagram of the channel configuration used in the experimental measurements*

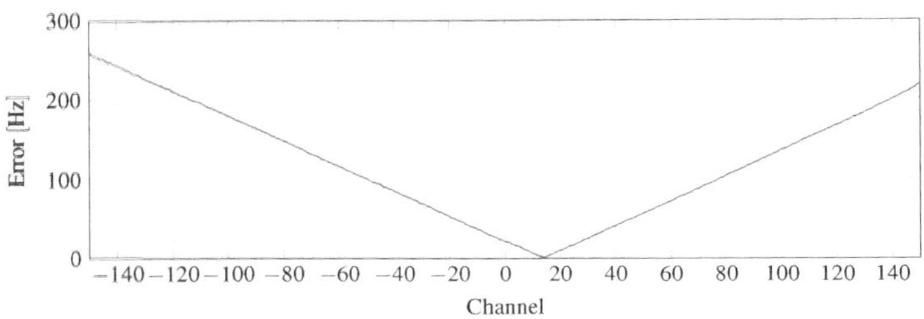

Figure 7.16. *Frequency deviation from the tag modulation frequency (-15 to 15 kHz) at each channel*

Figure 7.16 shows the frequency error at each channel, showing an increase in error as the frequency moves away from the central frequency (channel 0 at 300 kHz). Figure 7.17 shows the channel determined by the receiver before and after compensating for the frequency error introduced by the receiver.

It is important to clarify that the application proposed in this chapter (a wireless security system) does not need to address issues related to simultaneous communications from multiple sensors, as required by other applications. This is because an intrusion attempt typically targets a single specific entry point. In the unlikely event that access is attempted through multiple points simultaneously, the probability of triggering the sensors at the exact same moment is particularly low. Nevertheless, assuming the worst-case scenario where two or more sensors are activated at the same instant, it is sufficient to detect part of the preamble from one of the packets (presumably the one closest to the transmitter/receiver) to trigger the alarm. Therefore, the simultaneous activation and backscattering of multiple packets at the same time do not pose any issue.

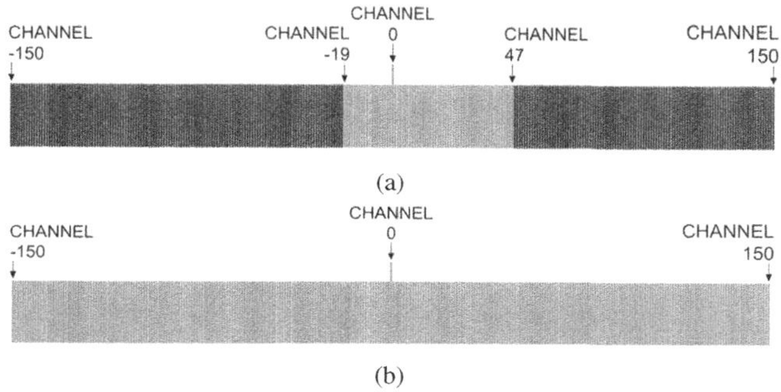

Figure 7.17. *Channel correct/incorrect determination before (a) and after (b) compensating for the frequency deviation introduced by the LoRa receiver*

Distance measurements are carried out to validate the capability of the proposed system and the range enhancement of the reflection amplifier compared to a non-amplified backscattering modulator. The measurements are carried out placing the tag 2.4 meters from the transmitter and moving the receiver away in a straight line.

The measurements are conducted in a hallway with a direct line of sight. It is important to note that this environment is not ideal for wave propagation, as waves do not travel freely without obstacles; rather, numerous interferences are caused by the floor, walls and railing, which create intermediate points where the signal is lost (Hashemi 1993; Molisch 2012). The transmission power is set at 10 dBm. Figure 7.18 shows the received signal strength (RSSI) as a function of distance.

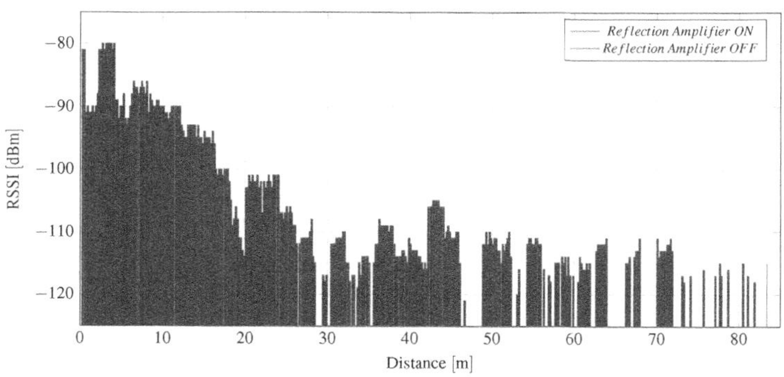

Figure 7.18. *Received signal strength indicator (RSSI) as a function of distance*

Finally, the system is validated in a real-case scenario, demonstrating its capability to cover a large household (150 m^2) with a single transmitter and receiver, as shown in Figure 7.19. LoRa transceivers are placed 4.5 meters apart, and critical points for potential external access (main entrance and windows) are evaluated by placing the wireless sensor at these locations. A total of nine points (P1–P9), shown in Figure 7.19, are tested.

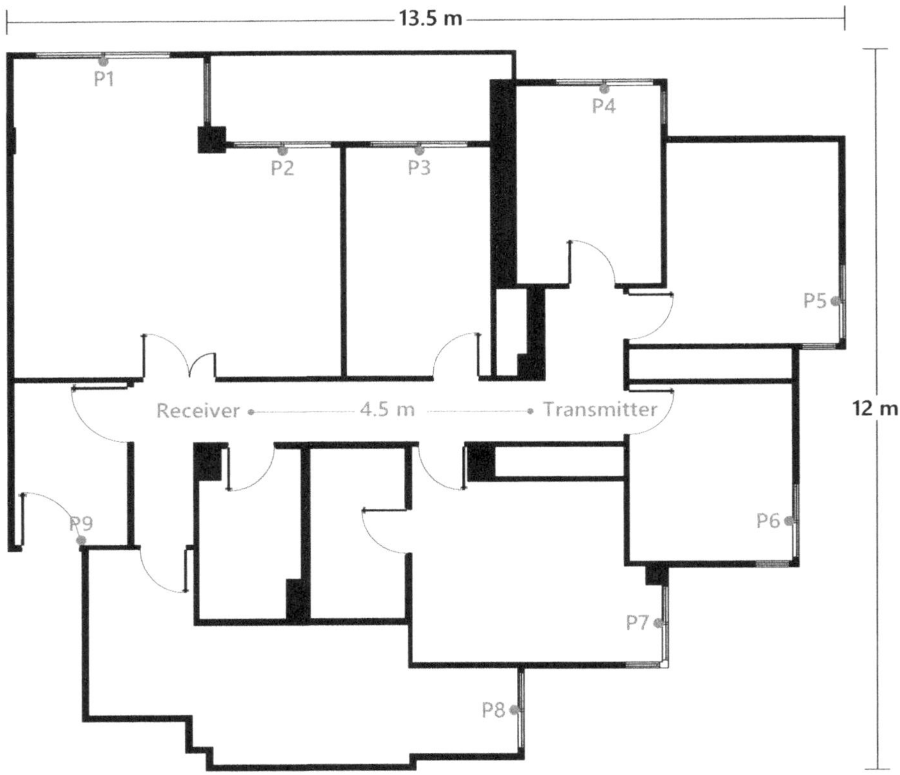

Figure 7.19. *Layout of the apartment where measurements are conducted, showing the positions of the transceivers and the assessment points for the wireless sensor (P1–P9)*

The system is evaluated at two transmission power levels (10 dBm and 0 dBm) with the reflection amplifier enabled and disabled, measuring the RSSI and SNR of the received packets. The measurement results are provided in Table 7.3. From the table, it can be observed that P4 and P5 are the most critical points despite not being the furthest from the transmitter. This measurement demonstrates that, even when the distance is within the system's maximum coverage range, environmental factors do not guarantee sensor coverage. Furthermore, the measurements show how the reflection amplifier resolves the coverage failure at points P4, P5 and P8. It can also be noted from the table's results that the reflection amplifier is more effective when the transmission power is lower or when the wireless nodes are farther from the transmitter. This behavior is known and supported by the amplifier characterization in the laboratory, where it was observed that the gain decreases as the input power increases, a consequence of the amplifier's gain compression.

Tx Power	Reflection Amplifier	Measure	Position								
			P1	P2	P3	P4	P5	P6	P7	P8	P9
0 dBm	Enabled	RSSI	-100.11	-101.09	-90.52	-106.8	-115.36	-93.62	-98.49	-107.57	-98.01
	Disabled		-116.27	-115.12	-101.20	-	-	-106.67	-107.55	-	-110.36
	Enabled	SNR	-2.04	-2.56	2.58	-7.01	-15.69	1.96	-0.98	-8.11	-0.93
	Disabled		-16.45	-15.16	-2.44	-	-	-7.08	-7.95	-	-10.56
10 dBm	Enabled	RSSI	-93.56	-92.02	-82.9	-101.0	-98.15	-88.62	-86.52	-96.08	-93.0
	Disabled		-105.53	-101.32	-86.69	-	-111.53	-99.98	-102.09	-105.02	-100.23
	Enabled	SNR	-8.19	-14.84	1.39	-9.17	-5.84	-11.85	-0.19	-5.72	-2.2
	Disabled		-12.42	-8.02	1.15	-	-19.6	-7.07	-9.15	-11.85	-7.25

Table 7.3. *Measurements of RSSI and SNR in the apartment depicted in Figure 7.19*

It is important to mention that the transmitter's maximum power is 22 dBm. In Spain, as stipulated by the Spanish Standardization Agency (UNE) in the UNE-EN 300220-2 standard (ETSI 2018), and similarly to other European countries, transmissions in the 433 ISM band cannot exceed 10 dBm. However, in other countries, these regulations may vary and can be more lenient, permitting higher power levels. Consequently, the system's final configuration strongly depends on local regulations and the installation environment. For example, in the apartment depicted in Figure 7.19, higher transmission power levels could allow for deactivating the amplifier across all wireless sensors, significantly reducing their energy consumption.

7.4. Conclusion

The work presented in this chapter describes the design of a surveillance system based on LoRa backscattering that improves several aspects of existing commercial systems. First, the power consumption of the wireless nodes is reduced by replacing conventional front-ends with a backscatter-based front-end. This reduction in power consumption allows the use of a supercapacitor recharged with solar energy instead of a typical button or LiPo batteries. Additionally, the ASK modulation used in commercial systems is replaced with LoRa modulation, making the system more resistant to interference and increasing the range of the backscattered signal. Second, a reflection amplifier with a phase modulator is designed and integrated in the backscattering front-end to increase the reflected power and consequently enhance the range. To identify which sensor triggers the alarm, a novel detection protocol is proposed based on the inherent ability of LoRa transceivers to determine the frequency deviation. The modulation is entirely analog, ensuring maximum efficiency and responsiveness compared to other digital solutions. The reliability of this communication protocol is experimentally validated by backscattering LoRa packets across 301 different channels, each separated by 100 Hz.

Finally, the system is validated in indoor environments, including both laboratory settings and a real-world scenario. Low-cost commercial LoRa transceivers are used as a cost-effective alternative to the expensive SDR equipment frequently cited in the literature. Measurements in the laboratory, conducted in a corridor with considerable multipath interference, demonstrate the range enhancement achieved by integrating the reflection amplifier, achieving a maximum of 84 meters with the reflection amplifier enabled, compared to 60 meters when it is disabled. The system is also tested in an apartment with two different transmission power levels, recording the RSSI and SNR metrics of the received packets. These results highlight the importance of the amplifier in providing coverage to the most challenging points and validate the system's operation in a real application. None of the measurements in the laboratory or in the apartment exceeded a transmission power of 10 dBm. It can be concluded that the range in other scenarios can be further enhanced by increasing transmission power (when permitted by regulations), deploying additional LoRa transceivers, strategically positioning them to mitigate fading caused by multipath interference or adjusting the CSS modulation parameters.

References

3rd Generation Partnership Project (2020a). 3GPP TS 36.211: Evolved universal terrestrial radio access (E-UTRA): Physical channels and modulation. Technical Specification, 3rd Generation Partnership Project, Sophia Antipolis [Online]. Available at: https://www.3gpp.org/ftp/Specs/archive/36_series/36.211/ [Accessed 25 June 2023].

3rd Generation Partnership Project (2020b). 3GPP TS 36.300: Evolved universal terrestrial radio access (E-UTRA) and evolved packet core (EPC): User equipment (UE) and mobile station (MS) radio access capabilities. Technical Specification, 3rd Generation Partnership Project, Sophia Antipolis [Online]. Available at: https://www.3gpp.org/ftp/Specs/archive/36_series/36.300/ [Accessed 25 June 2023].

Abedi, A., Mazaheri, M.H., Abari, O., Brecht, T. (2018). WiTAG: Rethinking backscatter communication for WiFi networks. In *Proceedings of the 17th ACM Workshop on Hot Topics in Networks*, New York.

Abedi, A., Dehbashi, F., Mazaheri, M.H., Abari, O., Brecht, T. (2020). WiTAG: Seamless WiFi backscatter communication. In *Proceedings of the Annual Conference of the ACM Special Interest Group on Data Communication on the Applications, Technologies, Architectures, and Protocols for Computer Communication*, New York.

Adeyeye, A., Lynch, C., Eid, A., Hester, J., Tentzeris, M. (2021). 5.8-GHz low-power tunnel-diode-based two-way repeater for non-line-of-sight interrogation of RFIDs and wireless sensor networks. *IEEE Microwave and Wireless Components Letters*, 31(6), 794–797.

Aernouts, M., Berkvens, R., Van Vlaenderen, K., Weyn, M. (2018). Sigfox and LoRaWAN datasets for fingerprint localization in large urban and rural areas. *Data*, 3(2), 13.

Ahson, S.A. and Ilyas, M. (2017). *RFID Handbook: Applications, Technology, Security, and Privacy*. CRC Press, Boca Raton.

Aitenbichler, E. and Muhlhauser, M. (2003). An IR local positioning system for smart items and devices. In *23rd International Conference on Distributed Computing Systems Workshops, 2003. Proceedings*. IEEE, Providence, RI.

Al-Sarawi, S., Anbar, M., Abdullah, R., Al Hawari, A.B. (2020). Internet of things market analysis forecasts, 2020–2030. In *2020 Fourth World Conference on Smart Trends in Systems, Security and Sustainability (WorldS4)*. IEEE, London.

Ali-Rantala, P., Ukkonen, L., Sydanheimo, L., Keskilammi, M., Kivikoski, M. (2003). Different kinds of walls and their effect on the attenuation of radiowaves indoors. In *IEEE Antennas and Propagation Society International Symposium*, Columbus, OH.

Alkhawaja, F., Jaradat, M., Romdhane, L. (2019). Techniques of indoor positioning systems (IPS): A survey. In *2019 Advances in Science and Engineering Technology International Conferences (ASET)*. IEEE, Dubai.

Alsharif, M.H., Kim, S., Kuruoğlu, N. (2019). Energy harvesting techniques for wireless sensor networks/radio-frequency identification: A review. *Symmetry*, 11(7), 865.

Amato, F., Peterson, C.W., Degnan, B.P., Durgin, G.D. (2015). A 45 μW bias power, 34 dB gain reflection amplifier exploiting the tunneling effect for RFID applications. In *2015 IEEE International Conference on RFID (RFID)*, San Diego, CA.

Amato, F., Peterson, C.W., Degnan, B.P., Durgin, G.D. (2018a). Tunneling RFID tags for long-range and low-power microwave applications. *IEEE Journal of Radio Frequency Identification*, 2(2), 93–103.

Amato, F., Torun, H.M., Durgin, G.D. (2018b). RFID backscattering in long-range scenarios. *IEEE Transactions on Wireless Communications*, 17(4), 2718–2725.

Analog Devices (2003). ADG901/ADG902 0 Hz to 4.5 GHz, 40 dB Off Isolation at 1 GHz,17 dBm P1dB at 1 GHz SPST Switches. Data Sheet, ADG902.

Analog Devices (2014). MAX9914–MAX9917: 1MHz, 20μA, Rail-to-Rail I/O Op Amps with Shutdown. Data Sheet, MAX9914–MAX9917.

Analog Devices (2016). ADG918/ADG919: Wideband 4 GHz, 43 dB Isolation at 1 GHz, CMOS 1.65 V to 2.75 V, 2:1 Mux/SPDT. Data Sheet, ADG918/ADG919.

Analog Devices (2021). MAX9634: nanoPower, 4-Bump UCSP/SOT23, Precision Current-Sense Amplifier. Data Sheet, MAX9634.

Andrade, C.B. and Hoefel, R.P.F. (2010). IEEE 802.11 WLANs: A comparison on indoor coverage models. In *CCECE 2010*. IEEE, Calgary, AB.

Andrushchak, V., Maksymyuk, T., Klymash, M., Ageyev, D. (2018). Development of the iBeacon's positioning algorithm for indoor scenarios. In *2018 International Scientific-Practical Conference Problems of Infocommunications. Science and Technology (PIC S&T)*. IEEE, Kharkiv.

Atmel (2013). ATtiny25/45/85/V: Atmel 8-bit AVR Microcontroller with 2/4/8K Bytes In-System Programmable Flash. Data Sheet, ATtiny25/V/ATtiny45/V/ATtiny85/V.

Atmel (2020). ATtiny417/814/816/817 tinyAVR 1-series. MCU de 8 Bits, AVR Family ATtiny417 Series Microcontrollers, AVR, 20 MHz, 4 KB, 24 Pines, VQFN. Data Sheet, ATtiny417/ATtiny814/ATtiny816/ATtiny817.

Atmel (2021). ATtiny202/204/402/404/406 tinyAVR 0-series. MCU de 8 Bits, AVR Family ATtiny402 Series Microcontrollers, AVR, 20 MHz, 4 KB, 8 Pines, SOIC. Data Sheet, ATtiny202/ATtiny204/ATtiny402/ATtiny404/ATtiny406.

Ayappa, K., Davis, H., Crapiste, G., Davis, E., Gordon, J. (1991). Microwave heating: An evaluation of power formulations. *Chemical Engineering Science*, 46(4), 1005–1016.

Babaeian, F. and Karmakar, N.C. (2018). Hybrid chipless RFID Tags – A pathway to EPC global standard. *IEEE Access*, 6, 67415–67426.

Bacon, F. (1859). *Meditationes Sacrae*. Excusum impensis Humfredi Hooper, London.

Bae, J., Song, K., Lee, H., Cho, H., Yoo, H.-J. (2011). A 0.24-nJ/b wireless body-area-network transceiver with scalable double-FSK modulation. *IEEE Journal of Solid-State Circuits*, 47(1), 310–322.

Bagchi, S., Abdelzaher, T.F., Govindan, R., Shenoy, P., Atrey, A., Ghosh, P., Xu, R. (2020). New frontiers in IoT: Networking, systems, reliability, and security challenges. *IEEE Internet of Things Journal*, 7(12), 11330–11346.

Balanis, C.A. (2016). *Antenna Theory: Analysis and Design*. John Wiley & Sons, New York.

Barba, A.B., Bianco, G.M., Fiore, L., Arduini, F., Marrocco, G., Occhiuzzi, C. (2022). Design and manufacture of flexible epidermal NFC device for electrochemical sensing of sweat. In *2022 IEEE International Conference on Flexible and Printable Sensors and Systems (FLEPS)*, Vienna.

Basiouny, Y., Arafa, M., Sarhan, A.M. (2017). Enhancing Wi-Fi fingerprinting for indoor positioning system using single multiplicative neuron and PCA algorithm. In *2017 12th International Conference on Computer Engineering and Systems (ICCES)*. IEEE, Cairo.

Belo, D., Correia, R., Ding, Y., Daskalakis, S.N., Goussetis, G., Georgiadis, A., Carvalho, N.B. (2019). IQ impedance modulator front-end for low-power LoRa backscattering devices. *IEEE Transactions on Microwave Theory and Techniques*, 67(12), 5307–5314.

Besnoff, J.S. and Reynolds, M.S. (2012). Near field modulated backscatter for in vivo biotelemetry. In *2012 IEEE International Conference on RFID (RFID)*, Orlando, FL.

Bharadia, D., Joshi, K.R., Kotaru, M., Katti, S. (2015). Backfi: High throughput WiFi backscatter. *ACM SIGCOMM Computer Communication Review*, 45(4), 283–296.

Blackman, S., Matlo, C., Bobrovitskiy, C., Waldoch, A., Fang, M.L., Jackson, P., Mihailidis, A., Nygård, L., Astell, A., Sixsmith, A. (2016). Ambient assisted living technologies for aging well: A scoping review. *Journal of Intelligent Systems*, 25(1), 55–69.

Bletsas, A., Dimitriou, A.G., Sahalos, J.N. (2010). Improving backscatter radio tag efficiency. *IEEE Transactions on Microwave Theory and Techniques*, 58(6), 1502–1509.

Bluetooth Special Interest Group (2020). Bluetooth core specification version 5.2 [Online]. Available at: https://www.bluetooth.com/specifications/specs/ [Accessed 25 June 2023].

Blumenthal, J., Grossmann, R., Golatowski, F., Timmermann, D. (2007). Weighted centroid localization in zigbee-based sensor networks. In *2007 IEEE International Symposium on Intelligent Signal Processing*, Alcala de Henares.

Boada, M., Lazaro, A., Villarino, R., Gil-Dolcet, E., Girbau, D. (2021). Battery-less NFC bicycle tire pressure sensor based on a force-sensing resistor. *IEEE Access*, 9, 103975–103987.

Boada, M., Lazaro, A., Girbau, D., Villarino, R. (2022). *Battery-less NFC Sensors for the Internet of Things*. ISTE Ltd, London and John Wiley & Sons, New York.

Brodsky, I. (2008). How Reginald Fessenden put wireless on the right technological footing. In *IEEE GLOBECOM 2008–2008, IEEE Global Telecommunications Conference*, New Orleans, LA.

Buoite Stella, A., Manganotti, P., Furlanis, G., Accardo, A., Ajčević, M. (2020). Return to school in the COVID-19 era: Considerations for temperature measurement. *Journal of Medical Engineering & Technology*, 44(8), 468–471.

Burdette, E.C., Cain, F.L., Seals, J. (1980). In vivo probe measurement technique for determining dielectric properties at VHF through microwave frequencies. *IEEE Transactions on Microwave Theory and Techniques*, 28(4), 414–427.

Čabarkapa, D., Grujić, I., Pavlović, P. (2015). Comparative analysis of the bluetooth low-energy indoor positioning systems. In *2015 12th International Conference on Telecommunication in Modern Satellite, Cable and Broadcasting Services (TELSIKS)*. IEEE, Nis.

Camera, F., Miozzi, C., Amato, F., Occhiuzzi, C., Marrocco, G. (2020). Experimental assessment of wireless monitoring of Axilla temperature by means of epidermal battery-less rfid sensors. *IEEE Sensors Letters*, 4(11), 1–4.

Cassel, M., Dépret, T., Piégay, H. (2017). Assessment of a new solution for tracking pebbles in rivers based on active RFID. *Earth Surface Processes and Landforms*, 42(13), 1938–1951.

Castelló-Palacios, S., Vallés-Lluch, A., Garcia-Pardo, C., Fornes-Leal, A., Cardona, N. (2017). Formulas for easy-to-prepare tailored phantoms at 2.4 GHz ISM band. In *2017 11th International Symposium on Medical Information and Communication Technology (ISMICT)*. IEEE, Lisbon.

Castelló-Palacios, S., Garcia-Pardo, C., Alloza-Pascual, M., Fornes-Leal, A., Cardona, N., Vallés-Lluch, A. (2019). Gel phantoms for body microwave propagation in the (2 to 26.5) GHz frequency band. *IEEE Transactions on Antennas and Propagation*, 67(10), 6564–6573.

Center, N.A.W. (1997). *Electronic Warfare and Radar Systems Engineering Handbook*. Electronic Warfare Division, Pont Mugu, CA.

Chan, P. and Fusco, V. (2013). Full duplex reflection amplifier tag. *IET Microwaves, Antennas & Propagation*, 7(6), 415–420.

Chen, H., Li, F., Wang, Y. (2016). EchoLoc: Accurate device-free hand localization using COTS devices. In *2016 45th International Conference on Parallel Processing (ICPP)*. IEEE, Philadelphia, PA.

Chen, H.-Y., Chen, A., Chen, C. (2020). Investigation of the impact of infrared sensors on core body temperature monitoring by comparing measurement sites. *Sensors*, 20(10), 2885.

Cheng, Y., Wang, C., Zhong, J., Lin, S., Xiao, Y., Zhong, Q., Jiang, H., Wu, N., Li, W., Chen, S. et al. (2017). Electrospun polyetherimide electret nonwoven for bi-functional smart face mask. *Nano Energy*, 34, 562–569.

Chi, Z., Liu, X., Wang, W., Yao, Y., Zhu, T. (2020). Leveraging ambient LTE traffic for ubiquitous passive communication. In *Proceedings of the Annual Conference of the ACM Special Interest Group on Data Communication on the Applications, Technologies, Architectures, and Protocols for Computer Communication*, New York.

Cho, H., Kim, H., Kim, M., Jang, J., Lee, Y., Lee, K.J., Bae, J., Yoo, H.-J. (2015). A 79 pJ/b 80 Mb/s Full-duplex transceiver and a 42.5μW 100 kb/s super-regenerative transceiver for body channel communication. *IEEE Journal of Solid-State Circuits*, 51(1), 310–317.

Choi, S.H. and Kim, D.I. (2015). Backscatter radio communication for wireless powered communication networks. In *2015 21st Asia-Pacific Conference on Communications (APCC)*. IEEE, Kyoto.

Chou, C.-K., Chen, G.-W., Guy, A.W., Luk, K.H. (1984). Formulas for preparing phantom muscle tissue at various radiofrequencies. *Bioelectromagnetics: Journal of the Bioelectromagnetics Society*, 5(4), 435–441.

Cichon, D.J. and Wiesbeck, W. (1995). The Heinrich Hertz wireless experiments at Karlsruhe in the view of modern communication. In *Proceedings of the 1995 International Conference on 100 Years of Radio*. IEEE, London.

Cnaan-On, I., Thomas, S.J., Reynolds, M.S., Krolik, J.L. (2014). Multichannel radar backscatter communication and localization. In *2014 IEEE International Conference on Acoustics, Speech and Signal Processing (ICASSP)*, Florence.

Cnaan-On, I., Thomas, S.J., Krolik, J.L., Reynolds, M.S. (2015). Multichannel backscatter communication and ranging for distributed sensing with an FMCW radar. *IEEE Transactions on Microwave Theory and Techniques*, 63(7), 2375–2383.

Cobos, M., Antonacci, F., Alexandridis, A., Mouchtaris, A., Lee, B. (2017). A survey of sound source localization methods in wireless acoustic sensor networks. *Wireless Communications and Mobile Computing*. doi: 10.1155/2017/3956282.

Costa, F., Genovesi, S., Borgese, M., Michel, A., Dicandia, F.A., Manara, G. (2021). A review of RFID sensors, the new frontier of internet of things. *Sensors*, 21(9), 3138.

Cui, S., Sun, B., Sun, X. (2020). A method for improving temperature measurement accuracy on an infrared thermometer for the ambient temperature field. *Review of Scientific Instruments*, 91(5), 054903

Dardari, D., Conti, A., Ferner, U., Giorgetti, A., Win, M.Z. (2009). Ranging with ultrawide bandwidth signals in multipath environments. *Proceedings of the IEEE*, 97(2), 404–426.

Daskalakis, S.-N., Kimionis, J., Collado, A., Tentzeris, M.M., Georgiadis, A. (2017a). Ambient FM backscattering for smart agricultural monitoring. In *2017 IEEE MTT-S International Microwave Symposium (IMS)*, Honololu, HI.

Daskalakis, S.N., Kimionis, J., Collado, A., Goussetis, G., Tentzeris, M.M., Georgiadis, A. (2017b). Ambient backscatterers using FM broadcasting for low cost and low power wireless applications. *IEEE Transactions on Microwave Theory and Techniques*, 65(12), 5251–5262.

Debes, C., Merentitis, A., Sukhanov, S., Niessen, M., Frangiadakis, N., Bauer, A. (2016). Monitoring activities of daily living in smart homes: Understanding human behavior. *IEEE Signal Processing Magazine*, 33(2), 81–94.

Del Mundo, L.B., Ansay, R.L.D., Festin, C.A.M., Ocampo, R.M. (2011). A comparison of wireless fidelity (Wi-Fi) fingerprinting techniques. In *ICTC 2011*. IEEE, Seoul.

Derévianckine, G.H., Guitton, A., Iova, O., Ning, B., Valois, F. (2023). Opportunities and challenges of LoRa 2.4 GHz. *IEEE Communications Magazine*, 61(10), 164–170.

DiCarlofelice, A., DiGiampaolo, E., Tognolatti, P. (2020). mm-Wave chipless RFID tag for healthcare applications. In *2020 XXXIIIrd General Assembly and Scientific Symposium of the International Union of Radio Science*. IEEE, Rome.

Dinh, T.-M.T., Duong, N.-S., Sandrasegaran, K. (2020). Smartphone-based indoor positioning using BLE iBeacon and reliable lightweight fingerprint map. *IEEE Sensors Journal*, 20(17), 10283–10294.

Djosic, S., Stojanovic, I., Jovanovic, M., Nikolic, T., Djordjevic, G.L. (2021). Fingerprinting-assisted UWB-based localization technique for complex indoor environments. *Expert Systems with Applications*, 167, 114188.

Donaldson, P. and Sayer, E. (1981). A technology for implantable hermetic packages. Part 2: An implementation. *Medical and Biological Engineering and Computing*, 19, 403–405.

Dong, W.-h., Wang, L., Yu, G.-z., Mei, Z.-b. (2016). Design of wireless automatic fire alarm system. *Procedia Engineering*, 135, 413–417.

Dunna, M., Meng, M., Wang, P.-H., Zhang, C., Mercier, P.P., Bharadia, D. (2021). SyncScatter: Enabling WiFi like synchronization and range for WiFi backscatter communication. In *Proceedings of the 18th USENIX Symposium on Networked Systems Design and Implementation*, Berkeley, CA.

Dylla, H. and Corneliussen, S.T. (2005). John Ambrose Fleming and the beginning of electronics. *Journal of Vacuum Science & Technology A: Vacuum, Surfaces, and Films*, 23(4), 1244–1251.

Eid, A., Hester, J., Tentzeris, M.M. (2020). A 5.8 GHz fully-tunnel-diodes-based 20 μW, 88mV, and 48 dB-gain fully-passive backscattering RFID Tag. In *2020 IEEE MTT-S International Microwave Symposium (IMS)*, Los Angeles, CA.

El Alami, H. and Najid, A. (2020). Optimization of energy efficiency in wireless sensor networks and internet of things: A review of related works. In *Nature-Inspired Computing Applications in Advanced Communication Networks*, Gupta, G.P. (ed.). IGI Global Scientific Publishing, Hershey, PA.

Elshabrawy, T. and Robert, J. (2018). The impact of ISM interference on LoRa BER performance. In *2018 IEEE Global Conference on Internet of Things (GCIoT)*, Alexandria.

Ensworth, J.F. and Reynolds, M.S. (2015). Every smart phone is a backscatter reader: Modulated backscatter compatibility with bluetooth 4.0 low energy (BLE) devices. In *2015 IEEE International Conference on RFID*, San Diego, CA.

ETSI (2018). ETSI EN 300 220-2 V3. 2.1: Short range devices (SRD) operating in the frequency range 25 MHz to 1000 MHz. Part 2: Harmonised standard for access to radio spectrum for non specific radio equipment. Harmonised European Standard, ETSI, Sophia Antipolis.

Faragher, R. and Harle, R. (2015). Location fingerprinting with bluetooth low energy beacons. *IEEE Journal on Selected Areas in Communications*, 33(11), 2418–2428.

Farzami, F., Khaledian, S., Smida, B., Erricolo, D. (2017a). Reconfigurable dual-band bidirectional reflection amplifier with applications in Van Atta array. *IEEE Transactions on Microwave Theory and Techniques*, 65(11), 4198–4207.

Farzami, F., Khaledian, S., Smida, B., Erricolo, D. (2017b). Ultra-low power reflection amplifier using tunnel diode for RFID applications. In *2017 IEEE International Symposium on Antennas and Propagation & USNC/URSI National Radio Science Meeting*, San Diego, CA.

Fasarakis-Hilliard, N., Alevizos, P.N., Bletsas, A. (2015). Coherent detection and channel coding for bistatic scatter radio sensor networking. *IEEE Transactions on Communications*, 63(5), 1798–1810.

Feng, J., Zhou, C., He, C., Li, Y., Ye, X. (2017). Development of an improved wearable device for core body temperature monitoring based on the dual heat flux principle. *Physiological Measurement*, 38(4), 652.

Finkenzeller, K. (2010). *RFID Handbook: Fundamentals and Applications in Contactless Smart Cards, Radio Frequency Identification and Near-Field Communication*. John Wiley & Sons, New York.

Firouzi, F., Farahani, B., Weinberger, M., DePace, G., Aliee, F.S. (2020). IoT fundamentals: Definitions, architectures, challenges, and promises. In *Intelligent Internet of Things: From Device to Fog and Cloud*, Firouzi, F., Chakrabarty, K., Nassif, S. (eds.) Springer, Cham.

Fischer, H.F., Wittmann, D., Baucells Costa, A., Zhou, B., Joost, G., Lukowicz, P. (2021). Masquare: A functional smart mask design for health monitoring. In *2021 International Symposium on Wearable Computers*. ACM, New York.

Fois, A., Tocco. F., Dell'Osa, A., Melis, L., Bertelli, U., Concu, A., Bertetto, A.M., Serra, C. (2021). Innovative smart face mask to protect workers from COVID-19 infection. In *2021 IEEE International Symposium on Medical Measurements and Applications (MeMeA)*, Lausanne.

Fong, S.L., Bakar, A.A.A.b.A., Ahmed, F.Y., Jamal, A. (2019). Smart transportation system using RFID. In *Proceedings of the 2019 8th International Conference on Software and Computer Applications*. ACM, New York.

Forouzandeh, M. and Karmakar, N.C. (2015). Chipless RFID tags and sensors: A review on time-domain techniques. *Wireless Power Transfer*, 2(2), 62–77.

Fox, R., Solman, A., Isaacs, R., Fry, A., MacDonald, I. (1973). A new method for monitoring deep body temperature from the skin surface. *Clinical Science and Molecular Medicine*, 44(1), 81–86.

Fu, X., Pedross-Engel, A., Arnitz, D., Watts, C.M., Sharma, A., Reynolds, M.S. (2017). Simultaneous imaging, sensor tag localization, and backscatter uplink via synthetic aperture radar. *IEEE Transactions on Microwave Theory and Techniques*, 66(3), 1570–1578.

Fulbrook, P. (1997). Core body temperature measurement: A comparison of axilla, tympanic membrane and pulmonary artery blood temperature. *Intensive and Critical Care Nursing*, 13(5), 266–272.

Gabriel, S., Lau, R., Gabriel, C. (1996). The dielectric properties of biological tissues: II. Measurements in the frequency range 10 Hz to 20 GHz. *Physics in Medicine & Biology*, 41(11), 2251.

Gao, Y., Cheng, S.-J., Toh, W.-D., Kwok, Y.-S., Tan, K.-C.B., Chen, X., Mok, W.-M., Win, H.-H., Zhao, B., Diao, S. et al. (2013). An asymmetrical QPSK/OOK transceiver SoC and 15: 1 JPEG encoder IC for multifunction wireless capsule endoscopy. *IEEE Journal of Solid-State Circuits*, 48(11), 2717–2733.

Getting, I.A. (1990). SCR-584 radar and the Mark 56 naval gun fire control system. *IEEE Aerospace and Electronic Systems Magazine*, 5(10), 3–15.

Ghatak, B., Banerjee, S., Ali, S.B., Bandyopadhyay, R., Das, N., Mandal, D., Tudu, B. (2020). Design of a self-powered smart mask for COVID-19. *arXiv:2005.08305*.

Girbau, D., Ramos, Á., Lazaro, A., Rima, S., Villarino, R. (2012). Passive wireless temperature sensor based on time-coded UWB chipless RFID tags. *IEEE Transactions on Microwave Theory and Techniques*, 60(11), 3623–3632.

Gogolak, L., Pletl, S., Kukolj, D. (2011). Indoor fingerprint localization in WSN environment based on neural network. In *2011 IEEE 9th International Symposium on Intelligent Systems and Informatics*, Subotica.

Goh, N.W.-J., Poh, J.-J., Yeo, J.Y., Aw, B.J.-J., Lai, S.C., Cheng, J.J.W., Tan, C.Y.L., Gan, S.K.-E. (2021). Design and development of a low cost, non-contact infrared thermometer with range compensation. *Sensors*, 21(11), 3817.

Goldsmith, A. (2005). *Wireless Communications*. Cambridge University Press, Cambridge.

Gong, W., Yuan, L., Wang, Q., Zhao, J. (2020). Multiprotocol backscatter for personal IoT sensors. In *Proceedings of the 16th International Conference on Emerging Networking EXperiments and Technologies*. ACM, New York.

Green, R.B. (1963). The general theory of antenna scattering. Thesis, Ohio State University, Columbus, OH.

Gu, J.-F., Wang, K., Wu, K. (2017). System architecture and signal processing for frequency-modulated continuous-wave radar using active backscatter tags. *IEEE Transactions on Signal Processing*, 66(9), 2258–2272.

Gumber, K., Dejous, C., Hemour, S. (2020). Harmonic reflection amplifier for widespread backscatter Internet-of-Things. *IEEE Transactions on Microwave Theory and Techniques*, 69(1), 774–785.

Gunga, H.-C., Sandsund, M., Reinertsen, R.E., Sattler, F., Koch, J. (2008). A non-invasive device to continuously determine heat strain in humans. *Journal of Thermal Biology*, 33(5), 297–307.

Guo, X., Shangguan, L., He, Y., Zhang, J., Jiang, H., Siddiqi, A.A., Liu, Y. (2020). Aloba: Rethinking ON-OFF keying modulation for ambient LoRa backscatter. In *Proceedings of the 18th Conference on Embedded Networked Sensor Systems*. ACM, New York.

Guo, X., Shangguan, L., He, Y., Zhang, J., Jiang, H., Siddiqi, A.A., Liu, Y. (2021). Efficient ambient LoRa backscatter with On-Off keying modulation. *IEEE/ACM Transactions on Networking*, 30(2), 641–654.

Guo, X., Shangguan, L., He, Y., Jing, N., Zhang, J., Jiang, H., Liu, Y. (2022). Saiyan: Design and implementation of a low-power demodulator for LoRa backscatter systems. In *19th USENIX Symposium on Networked Systems Design and Implementation*, Berkeley, CA.

Gupta, M., Abdelsalam, M., Mittal, S. (2020). Enabling and enforcing social distancing measures using smart city and its infrastructures: A COVID-19 use case. *arXiv:2004.09246.*

Gupta, D., Bhatt, S., Gupta, M., Tosun, A.S. (2021). Future smart connected communities to fight COVID-19 outbreak. *Internet of Things*, 13, 100342.

Ha, G.Y., Seo, S.B., Oh, H.S., Jeon, W.S. (2019). LoRa ToA-based localization using fingerprint method. In *2019 International Conference on Information and Communication Technology Convergence (ICTC)*. IEEE, Jeju.

Haddara, M. and Staaby, A. (2018). RFID applications and adoptions in healthcare: A review on patient safety. *Procedia Computer Science*, 138, 80–88.

Hall, R. (1960). Tunnel diodes. *IRE Transactions on Electron Devices*, 7(1), 1–9.

Hansen, R.C. (1989). Relationships between antennas as scatterers and as radiators. *Proceedings of the IEEE*, 77(5), 659–662.

Haritsa, T.R., Yashu, B., Kumar, U., Suma, M. (2020). Mathematical characterization and simulation of LoRa. *Wireless Personal Communications*, 115(2), 1481–1506.

Harrington, R.F. (1962). Small resonant scatterers and their use for field measurements. *IRE Transactions on Microwave Theory and Techniques*, 10(3), 165–174.

Harrington, R.F. (1964). Theory of loaded scatterers. *Proceedings of the Institution of Electrical Engineers*, 111(4), 617–623. doi: 10.1049/piee.1964.0111.

Hashemi, H. (1993). The indoor radio propagation channel. *Proceedings of the IEEE*, 81(7), 943–968.

He, Z., Cui, B., Zhou, W., Yokoi, S. (2015). A proposal of interaction system between visitor and collection in museum hall by iBeacon. In *2015 10th International Conference on Computer Science & Education (ICCSE)*. IEEE, Cambridge.

He, X., Jiang, W., Cheng, M., Zhou, X., Wan, P.-J., Yang, P., Kurkoski, B. (2019). GuardRider: Towards sustainable backscattering system over WiFi in the wild. *arXiv:1912.06493.*

He, X., Jiang, W., Cheng, M., Zhou, X., Yang, P., Kurkoski, B. (2020). Guardrider: Reliable WiFi backscatter using reed-Solomon codes with QoS guarantee. In *2020 IEEE/ACM 28th International Symposium on Quality of Service (IWQoS)*. IEEE, Hang Zhou.

Hessar, M., Najafi, A., Gollakota, S. (2019). Netscatter: Enabling large-scale backscatter networks. In *Proceedings of the 16th USENIX Symposium on Networked Systems Design and Implementation (NSDI'19)*. ACM, New York.

Hu, J., Zhong, L., Ma, T., Ding, Z., Xu, Z. (2021). Long-range FM backscatter tag with tunnel diode. *IEEE Microwave and Wireless Components Letters*, 32(1), 92–95.

Huang, F.-J., Lee, C.-M., Chang, C.-L., Chen, L.-K., Yo, T.-C., Luo, C.-H. (2011). Rectenna application of miniaturized implantable antenna design for triple-band biotelemetry communication. *IEEE Transactions on Antennas and Propagation*, 59(7), 2646–2653.

Huang, M., Tamura, T., Yoshimura, T., Tsuchikawa, T., Kanaya, S. (2016). Wearable deep body thermometers and their uses in continuous monitoring for daily healthcare. In *2016 38th Annual International Conference of the IEEE Engineering in Medicine and Biology Society (EMBC)*, Orlando, FL.

Ijaz, F., Yang, H.K., Ahmad, A.W., Lee, C. (2013). Indoor positioning: A review of indoor ultrasonic positioning systems. In *2013 15th International Conference on Advanced Communications Technology (ICACT)*. IEEE, Pyeongchang.

Ilsley, A., Rutten, A., Runciman, W. (1983). An evaluation of body temperature measurement. *Anaesthesia and Intensive Care*, 11(1), 31–39.

Infineon Technologies (2019). BFP405: Surface mount wideband silicon NPN RF bipolar transistor. Data Sheet, BFP405.

Ishizaki, H., Ikeda, H., Yoshida, Y., Maeda, T., Kuroda, T., Mizuno, M. (2011). A battery-less WiFi-BER modulated data transmitter with ambient radio-wave energy harvesting. In *2011 Symposium on VLSI Circuits-Digest of Technical Papers*. IEEE, Kyoto.

Islam, M.N. and Yuce, M.R. (2016). Review of medical implant communication system (MICS) band and network. *Ict Express*, 2(4), 188–194.

Iyer, V., Talla, V., Kellogg, B., Gollakota, S., Smith, J. (2016). Inter-technology backscatter: Towards internet connectivity for implanted devices. In *Proceedings of the 2016 ACM SIGCOMM Conference*, New York.

Jia, X. and Zhou, X. (2021). Power beacon placement for maximizing guaranteed coverage in bistatic backscatter networks. *IEEE Transactions on Communications*, 69(11), 7895–7909.

Jia, R., Jin, M., Chen, Z., Spanos, C.J. (2015). SoundLoc: Accurate room-level indoor localization using acoustic signatures. In *2015 IEEE International Conference on Automation Science and Engineering (CASE)*, Gothenburg.

Jia, X., Zhou, X., Niyato, D., Zhao, J. (2021). Intelligent reflecting surface-assisted bistatic backscatter networks: Joint beamforming and reflection design. *IEEE Transactions on Green Communications and Networking*, 6(2), 799–814.

Jiang, D., Cirmirakis, D., Schormans, M., Perkins, T.A., Donaldson, N., Demosthenous, A. (2016). An integrated passive phase-shift keying modulator for biomedical implants with power telemetry over a single inductive link. *IEEE Transactions on Biomedical Circuits and Systems*, 11(1), 64–77.

Jiang, Y., Pan, K., Leng, T., Hu, Z. (2019). Smart textile integrated wireless powered near field communication body temperature and sweat sensing system. *IEEE Journal of Electromagnetics, RF and Microwaves in Medicine and Biology*, 4(3), 164–170.

Jiang, J., Xu, Z., Dang, F., Wang, J. (2021a). Long-range ambient LoRa backscatter with parallel decoding. In *Proceedings of the 27th Annual International Conference on Mobile Computing and Networking*. ACM, New York.

Jiang, H., Zhang, J., Guo, X., He, Y. (2021b). Sense me on the ride: Accurate mobile sensing over a LoRa backscatter channel. In *Proceedings of the 19th ACM Conference on Embedded Networked Sensor Systems*, New York.

Jin, G.-y., Lu, X.-y., Park, M.-S. (2006). An indoor localization mechanism using active RFID tag. In *IEEE International Conference on Sensor Networks, Ubiquitous, and Trustworthy Computing (SUTC'06)*, Taichung.

Josephson, C., Barnhart, B., Katti, S., Winstein, K., Chandra, R. (2019). RF soil moisture sensing via radar backscatter tags. *arXiv:1912.12382*.

Josephson, C., Kotaru, M., Winstein, K., Katti, S., Chandra, R. (2021). Low-cost in-ground soil moisture sensing with radar backscatter tags. In *ACM SIGCAS Conference on Computing and Sustainable Societies*. ACM, New York.

Jung, J.-Y., Kim, H., Lee, H.-S., Yeom, K.-W. (2009). An UHF RFID tag with long read range. In *2009 European Microwave Conference (EuMC)*. IEEE, Rome.

Jung, J., Ryoo, J., Yi, Y., Kim, S.M. (2020). Gateway over the air: Towards pervasive internet connectivity for commodity IoT. In *Proceedings of the 18th International Conference on Mobile Systems, Applications, and Services*. ACM, New York.

Kahn, W. and Kurss, H. (1965). Minimum-scattering antennas. *IEEE Transactions on Antennas and Propagation*, 13(5), 671–675.

Kaiser, M.S., Mahmud, M., Noor, M.B.T., Zenia, N.Z., Mamun, S.A., Mahmud, K.M.A., Azad, S., Aradhya, V.N.M., Stephan, P., Stephan, T. et al. (2021). Worksafe: Towards healthy workplaces during COVID-19 with an intelligent Phealth app for industrial settings. *IEEE Access*, 9, 13814–13828.

Kalavakonda, R.R., Masna, N.V.R., Bhuniaroy, A., Mandal, S., Bhunia, S. (2020). A smart mask for active defense against Coronaviruses and other airborne pathogens. *IEEE Consumer Electronics Magazine*, 10(2), 72–79.

Kalavakonda, R.R., Masna, N.V.R., Mandal, S., Bhunia, S. (2021). A smart mask for active defense against airborne pathogens. *Scientific Reports*, 11(1), 1–9.

Kandel, L.N. and Yu, S. (2019). VWAN: Virtual WiFi ANtennas for increased indoor localization accuracy. In *2019 IEEE International Conference on Industrial Internet (ICII)*, Orlando, FL.

Kapetanovic, Z., Saffari, A., Chandra, R., Smith, J.R. (2019). Glaze: Overlaying occupied spectrum with downlink IoT transmissions. *Proceedings of the ACM on Interactive, Mobile, Wearable and Ubiquitous Technologies*, 3(4), 1–21.

Katanbaf, M., Weinand, A., Talla, V. (2021). Simplifying backscatter deployment: Full-duplex LoRa backscatter. In *NSDI*. USENIX, Berkeley, CA.

Kellogg, B., Parks, A., Gollakota, S., Smith, J.R., Wetherall, D. (2014). Wi-Fi backscatter: Internet connectivity for RF-powered devices. In *Proceedings of the 2014 ACM Conference on SIGCOMM*. New York.

Kellogg, B., Talla, V., Gollakota, S., Smith, J.R. (2016). Passive wi-fi: Bringing low power to wi-fi transmissions. In *13th Symposium on Networked Systems Design and Implementation*. USENIX, Berkeley, CA.

Khaledian, S., Farzami, F., Erricolo, D., Smida, B. (2017). A full-duplex bidirectional amplifier with low DC power consumption using tunnel diodes. *IEEE Microwave and Wireless Components Letters*, 27(12), 1125–1127.

Khaledian, S., Farzami, F., Smida, B., Erricolo, D. (2019). Two-way backscatter communication tag using a reflection amplifier. *IEEE Microwave and Wireless Components Letters*, 29(6), 421–423.

Khaleghi, A. and Balasingham, I. (2019). Wireless capsule endoscopy using backscatter communication. In *2019 13th European Conference on Antennas and Propagation (EuCAP)*. IEEE, Krakow.

Khaleghi, A., Hasanvand, A., Balasingham, I. (2018). Radio frequency backscatter communication for high data rate deep implants. *IEEE Transactions on Microwave Theory and Techniques*, 67(3), 1093–1106.

Kim, N., Wei, J.L.J., Ying, J., Zhang, H., Moon, S.K., Choi, J. (2020). A customized smart medical mask for healthcare personnel. In *2020 IEEE International Conference on Industrial Engineering and Engineering Management (IEEM)*, Singapore.

Kimionis, J., Bletsas, A., Sahalos, J.N. (2012). Bistatic backscatter radio for tag read-range extension. In *2012 IEEE International Conference on RFID-Technologies and Applications (RFID-TA)*, Nice.

Kimionis, J., Bletsas, A., Sahalos, J.N. (2013). Bistatic backscatter radio for power-limited sensor networks. In *2013 IEEE Global Communications Conference (GLOBECOM)*, Atlanta, GA.

Kimionis, J., Georgiadis, A., Collado, A., Tentzeris, M.M. (2014a). Enhancement of RF tag backscatter efficiency with low-power reflection amplifiers. *IEEE Transactions on Microwave Theory and Techniques*, 62(12), 3562–3571.

Kimionis, J., Tentzeris, M.M., Georgiadis, A., Collado, A. (2014b). Inkjet-printed reflection amplifier for increased-range backscatter radio. In *2014 44th European Microwave Conference*. IEEE, Rome.

Kimionis, J., Georgiadis, A., Kim, S., Collado, A., Niotaki, K., Tentzeris, M.M. (2014c). An enhanced-range RFID tag using an ambient energy powered reflection amplifier. In *2014 IEEE MTT-S International Microwave Symposium (IMS2014)*, Tampa, FL.

Kimionis, J., Bletsas, A., Sahalos, J.N. (2014d). Increased range bistatic scatter radio. *IEEE Transactions on Communications*, 62(3), 1091–1104.

Kiourti, A. and Nikita, K.S. (2012). A review of implantable patch antennas for biomedical telemetry: Challenges and solutions (wireless corner). *IEEE Antennas and Propagation Magazine*, 54(3), 210–228.

Kiourti, A., Christopoulou, M., Nikita, K.S. (2011). Performance of a novel miniature antenna implanted in the human head for wireless biotelemetry. In *2011 IEEE International Symposium on Antennas and Propagation (APSURSI)*, Spokane, WA.

Kitamura, K.-I., Zhu, X., Chen, W., Nemoto, T. (2010). Development of a new method for the noninvasive measurement of deep body temperature without a heater. *Medical Engineering and Physics*, 32(1), 1–6.

Koelle, A.R., Depp, S.W., Freyman, R.W. (1975). Short-range radio-telemetry for electronic identification, using modulated RF backscatter. *Proceedings of the IEEE*, 63(8), 1260–1261.

Konate, S., Li, C., Xu, L. (2020). LoRa backscatter automated irrigation approach: Reviewing and proposed system. In *2020 6th International Conference on Robotics and Artificial Intelligence*. ACM, New York.

Kramer, O. (ed.) (2016). Scikit-learn. In *Machine Learning for Evolution Strategies*. Springer, Cham.

Kraus, J.D., Marhefka, R.J., Khan, A.S. (2006). *Antennas and Wave Propagation*. McGraw Hill, New York.

Kumbhar, A. (2017). Overview of ism bands and software-defined radio experimentation. *Wireless Personal Communications*, 97(3), 3743–3756.

Kuo, Y.-S., Pannuto, P., Hsiao, K.-J., Dutta, P. (2014). Luxapose: Indoor positioning with mobile phones and visible light. In *Proceedings of the 20th Annual International Conference on Mobile Computing and Networking*. ACM, New York.

Lam, K.-H., Cheung, C.-C., Lee, W.-C. (2018). New RSSI-based LoRa localization algorithms for very noisy outdoor environment. In *2018 IEEE 42nd Annual Computer Software and Applications Conference (COMPSAC)*, Tokyo.

Lam, K.-H., Cheung, C.-C., Lee, W.-C. (2019). RSSI-based LoRa localization systems for large-scale indoor and outdoor environments. *IEEE Transactions on Vehicular Technology*, 68(12), 11778–11791.

Landt, J. (2005). The history of RFID. *IEEE Potentials*, 24(4), 8–11.

Lazaro, A., Girbau, D., Salinas, D. (2009). Radio link budgets for UHF RFID on multipath environments. *IEEE Transactions on Antennas and Propagation*, 57(4), 1241–1251.

Lazaro, A., Lorenzo, J., Villarino, R., Girbau, D. (2014). Backscatter transponder based on frequency selective surface for FMCW radar applications. *Radioengineering*, 23(2), 632–641.

Lazaro, A., Lorenzo, J., Villarino, R., Girbau, D. (2015). Modulated corner reflector using frequency selective surfaces for FMCW radar applications. In *2015 European Microwave Conference (EuMC)*. IEEE, Paris.

Lazaro, A., Villarino, R., Girbau, D. (2018). A survey of NFC sensors based on energy harvesting for IoT applications. *Sensors*, 18(11), 3746.

Lazaro, A., Boada, M., Villarino, R., Girbau, D. (2019). Study on the reading of energy-harvested implanted NFC tags using mobile phones. *IEEE Access*, 8, 2200–2221.

Lazaro, A., Lazaro, M., Villarino, R., De Paco, P. (2020a). New radar micro-Doppler tag for road safety based on the signature of rotating backscatters. *IEEE Sensors Journal*, 21(6), 8604–8612.

Lazaro, M., Lazaro, A., Villarino, R. (2020b). Feasibility of backscatter communication using LoRAWAN signals for deep implanted devices and wearable applications. *Sensors*, 20(21), 6342.

Lazaro, A., Lazaro, M., Villarino, R., Girbau, D., de Paco, P. (2021a). Car2car communication using a modulated backscatter and automotive fmcw radar. *Sensors*, 21(11), 3656.

Lazaro, M., Lazaro, A., Villarino, R., Gibau, D. (2021b). Smart mask for temperature monitoring with LoRa backscattering communication. In *2021 International Multidisciplinary Conference on Computer and Energy Science (SpliTech)*, Bol and Split.

Lazaro, A., Porcel, A., Lazaro, M., Villarino, R., Girbau, D. (2022a). Spoofing attacks on FMCW radars with low-cost backscatter tags. *Sensors*, 22(6), 2145.

Lazaro, A., Villarino, R., Lazaro, M., Canellas, N., Prieto-Simon, B., Girbau, D. (2022b). Battery-less NFC potentiostat for electrochemical point-of-care sensors based on COTS components. *Sensors*, 22(19), 7213.

Lazaro, M., Lazaro, A., González, B., Villarino, R., Girbau, D. (2022c). Long-range wireless system for U-value assessment using a low-cost heat flux sensor. *Sensors*, 22(19), 7259.

Lazaro, A., Lazaro, M., Villarino, R., Girbau, D. (2023). Smart spread spectrum modulated tags for detection of vulnerable road users with automotive radar. *Sensors*, 23(5), 2730.

Lee, J. and Yang, K. (2016). RF power analysis on 5.8 GHz low-power amplifier using resonant tunneling diodes. *IEEE Microwave and Wireless Components Letters*, 27(1), 61–63.

Lee, J., Lee, J., Yang, K. (2014). Reflection-type RTD low-power amplifier with deep sub-mW DC power consumption. *IEEE Microwave and Wireless Components Letters*, 24(8), 551–553.

Lefrant, J.-Y., Muller, L., de La Coussaye, J.E., Benbabaali, M., Lebris, C., Zeitoun, N., Mari, C., Saissi, G., Ripart, J., Eledjam, J.-J. (2003). Temperature measurement in intensive care patients: Comparison of urinary bladder, oesophageal, rectal, axillary, and inguinal methods versus pulmonary artery core method. *Intensive Care Medicine*, 29(3), 414–418.

Li, Y., He, Z., Li, Y., Xu, H., Pei, L., Zhang, Y. (2018). Towards location enhanced IoT: Characterization of LoRa signal for wide area localization. In *2018 Ubiquitous Positioning, Indoor Navigation and Location-Based Services (UPINLBS)*. IEEE, Wuhan.

Li, M., Zhao, L., Tan, D., Tong, X. (2019). BLE fingerprint indoor localization algorithm based on eight-neighborhood template matching. *Sensors*, 19(22), 4859.

Liao, J., Wang, X., Ruttik, K., Jantti, R., Dinh-Thuy, P.-H. (2023). Ambient FSK backscatter communications using LTE cell specific reference signals. *arXiv:2301.13664*.

Light, R.A. (2017). Mosquitto: Server and client implementation of the MQTT protocol. *Journal of Open Source Software*, 2(13), 265.

Lin, Y.-P., Yeh, C.-Y., Huang, P.-Y., Wang, Z.-Y., Cheng, H.-H., Li, Y.-T., Chuang, C.-F., Huang, P.-C., Tang, K.-T., Ma, H.-P. et al. (2015). A battery-less, implantable neuro-electronic interface for studying the mechanisms of deep brain stimulation in rat models. *IEEE Transactions on Biomedical Circuits and Systems*, 10(1), 98–112.

Lin, Y.-C., Sun, C.-C., Huang, K.-T. (2019). RSSI measurement with channel model estimating for IoT wide range localization using LoRa communication. In *2019 International Symposium on Intelligent Signal Processing and Communication Systems (ISPACS)*. IEEE, Taipei.

Linear Technologies (2005). LTC6907: Micropower, 40kHz to 4MHz Resistor Set Oscillator in SOT-23. Data Sheet, LTC6907.

Linear Technologies (2020). LTC6990: TimerBlox: Voltage Controlled Silicon Oscillator. Data Sheet, LTC6990.

Liu, W.-C., Yeh, F.-M., Ghavami, M. (2008). Miniaturized implantable broadband antenna for biotelemetry communication. *Microwave and Optical Technology Letters*, 50(9), 2407–2409.

Liu, V., Parks, A., Talla, V., Gollakota, S., Wetherall, D., Smith, J.R. (2013a). Ambient backscatter: Wireless communication out of thin air. *ACM SIGCOMM Computer Communication Review*, 43(4), 39–50.

Liu, Y.-H., Chen, L.-G., Lin, C.-Y., Lin, T.-H. (2013b). A 650-pJ/bit MedRadio transmitter with an FIR-embedded phase modulator for medical micro-power networks (MMNs). *IEEE Transactions on Circuits and Systems I: Regular Papers*, 60(12), 3279–3288.

Liu, X., Chi, Z., Wang, W., Yao, Y., Zhu, T. (2020a). VMscatter: A versatile MIMO backscatter. In *Proceedings of the 17th USENIX Symposium on Networked Systems Design and Implementation (NSDI '20)*, Berkeley, CA.

Liu, M., Cheng, L., Qian, K., Wang, J., Wang, J., Liu, Y. (2020b). Indoor acoustic localization: A survey. *Human-centric Computing and Information Sciences*, 10(2), 1–24.

Liu, X., Chi, Z., Wang, W., Yao, Y., Hao, P., Zhu, T. (2021). Verification and redesign of OFDM backscatter. In *Proceedings of the 18th USENIX Symposium on Networked Systems Design and Implementation*, Berkeley, CA.

Lorenzo, J., Lazaro, A., Villarino, R., Girbau, D. (2015). Active backscatter transponder for FMCW radar applications. *IEEE Antennas and Wireless Propagation Letters*, 14, 1610–1613.

Lorenzo, J., Lazaro, A., Girbau, D., Villarino, R., Gil, E. (2016). Analysis of on-body transponders based on frequency selective surfaces. *Progress in Electromagnetics Research*, 157, 133–143.

Lorenzo, J., Lazaro, A., Villarino, R., Girbau, D. (2017). Diversity study of a frequency selective surface transponder for wearable applications. *IEEE Transactions on Antennas and Propagation*, 65(5), 2701–2706.

Lu, S.-H. and Dai, Y.-T. (2009). Normal body temperature and the effects of age, sex, ambient temperature and body mass index on normal oral temperature: A prospective, comparative study. *International Journal of Nursing Studies*, 46(5), 661–668.

Ma, S., Sydänheimo, L., Ukkonen, L., Björninen, T. (2018). Split-ring resonator antenna system with cortical implant and head-worn parts for effective far-field implant communications. *IEEE Antennas and Wireless Propagation Letters*, 17(4), 710–713.

Martin, P., Ho, B.-J., Grupen, N., Munoz, S., Srivastava, M. (2014). An Ibeacon primer for indoor localization: Demo abstract. In *Proceedings of the 1st ACM Conference on Embedded Systems for Energy-Efficient Buildings*, New York.

Masna, N.V.R., Kalavakonda, R.R., Dizon, R., Bhunia, S. (2021). Smart and connected mask for protection beyond the pandemic. In *2021 IEEE International Midwest Symposium on Circuits and Systems (MWSCAS)*, Lansing, MI.

Mathews, I., King, P.J., Stafford, F., Frizzell, R. (2015). Performance of III–V solar cells as indoor light energy harvesters. *IEEE Journal of Photovoltaics*, 6(1), 230–235.

Matos, D., Correia, R., Carvalho, N.B. (2021). Dual-band FET-based reflection amplifier for backscatter modulator performance enhancement. *URSI Radio Science Letters*, 3, 39.

Maxwell, J.C. (1865). VIII. A dynamical theory of the electromagnetic field. *Philosophical Transactions of the Royal Society of London*, 155, 459–512.

Medina, C., Segura, J.C., De la Torre, A. (2013). Ultrasound indoor positioning system based on a low-power wireless sensor network providing sub-centimeter accuracy. *Sensors*, 13(3), 3501–3526.

Menon, R., Gujarathi, R., Saffari, A., Smith, J.R. (2022). Wireless identification and sensing platform version 6.0. In *SenSys '22: The 20th ACM Conference on Embedded Networked Sensor Systems*, New York.

Merli, F., Fuchs, B., Mosig, J.R., Skrivervik, A.K. (2010). The effect of insulating layers on the performance of implanted antennas. *IEEE Transactions on Antennas and Propagation*, 59(1), 21–31.

Microsemi (2015). ZL70103 Medical Implantable RF Transceiver. MICS-Band RF Telemetry. Data Sheet, ZL70103.

Milici, S., Lazaro, A., Lorenzo, J., Villarino, R., Girbau, D. (2017a). Wearable sensors based on modulated frequency selective surfaces. In *2017 47th European Microwave Conference (EuMC)*. IEEE, Nuremberg.

Milici, S., Lorenzo, J., Lázaro, A., Villarino, R., Girbau, D. (2017b). Wireless breathing sensor based on wearable modulated frequency selective surface. *IEEE Sensors Journal*, 17(5), 1285–1292.

Milici, S., Lázaro, A., Villarino, R., Girbau, D., Magnarosa, M. (2018). Wireless wearable magnetometer-based sensor for sleep quality monitoring. *IEEE Sensors Journal*, 18(5), 2145–2152.

Miller, D.J., Capodilupo, J.V., Lastella, M., Sargent, C., Roach, G.D., Lee, V.H., Capodilupo, E.R. (2020). Analyzing changes in respiratory rate to predict the risk of COVID-19 infection. *PLoS One*, 15(12), e0243693.

Miozzi, C., Amendola, S., Bergamini, A., Marrocco, G. (2017). Reliability of a re-usable wireless epidermal temperature sensor in real conditions. In *2017 IEEE 14th International Conference on Wearable and Implantable Body Sensor Networks (BSN)*, Eindhoven.

Miozzi, C., Saggio, G., Gruppioni, E., Marrocco, G. (2019). Constrained safety-integrity performance of through-the-arms UHF-RFID transcutaneous wireless communication for the control of prostheses. *IEEE Journal of Radio Frequency Identification*, 3(4), 236–244.

Miozzi, C., Amato, F., Marrocco, G. (2020). Performance and durability of thread antennas as stretchable epidermal UHF RFID tags. *IEEE Journal of Radio Frequency Identification*, 4(4), 398–405.

Mishra, D.P. and Behera, S.K. (2022). Resonator based chipless RFID: A frequency domain comprehensive review. *IEEE Transactions on Instrumentation and Measurement*, 72, 1–16, 5500716.

Molisch, A.F. (2012). *Wireless Communications*. John Wiley & Sons, New York.

Moravek, P., Komosny, D., Simek, M., Girbau, D., Lazaro, A. (2011). Energy analysis of received signal strength localization in wireless sensor networks. *Radioengineering*, 20(4), 937–944.

Moravek, P., Komosny, D., Simek, M., Jelinek, M., Girbau, D., Lazaro, A. (2013). Investigation of radio channel uncertainty in distance estimation in wireless sensor networks. *Telecommunication Systems*, 52, 1549–1558.

Murphy, O.H., McLeod, C.N., Navaratnarajah, M., Yacoub, M., Toumazou, C. (2011). A pseudo-normal-mode helical antenna for use with deeply implanted wireless sensors. *IEEE Transactions on Antennas and Propagation*, 60(2), 1135–1139.

Nemoto, T. and Togawa, T. (1988). Improved probe for a deep body thermometer. *Medical and Biological Engineering and Computing*, 26(4), 456–459.

Nessa, A., Adhikari, B., Hussain, F., Fernando, X.N. (2020). A survey of machine learning for indoor positioning. *IEEE Access*, 8, 214945–214965.

Nguyen, D.H., Zomorrodi, M., Karmakar, N.C. (2019). Spatial-based chipless RFID system. *IEEE Journal of Radio Frequency Identification*, 3(1), 46–55.

Ni, L.M., Liu, Y., Lau, Y.C., Patil, A.P. (2003). LANDMARC: Indoor location sensing using active RFID. In *Proceedings of the First IEEE International Conference on Pervasive Computing and Communications, 2003 (PerCom 2003)*, Fort Worth, TX.

Nikitin, P.V. and Rao, K.S. (2006). Theory and measurement of backscattering from RFID tags. *IEEE Antennas and Propagation Magazine*, 48(6), 212–218.

Nikitin, P.V., Rao, K., Martinez, R.D. (2007). Differential RCS of RFID tag. *Electronics Letters*, 43(8), 1.

Nikolayev, D., Zhadobov, M., Le Coq, L., Karban, P., Sauleau, R. (2017). Robust ultraminiature capsule antenna for ingestible and implantable applications. *IEEE Transactions on Antennas and Propagation*, 65(11), 6107–6119.

Noreen, U., Bounceur, A., Clavier, L. (2017). A study of LoRa low power and wide area network technology. In *2017 International Conference on Advanced Technologies for Signal and Image Processing (ATSIP)*. IEEE, Fez.

Nunes-Silva, P., Hrncir, M., Guimarães, J., Arruda, H., Costa, L., Pessin, G., Siqueira, J., De Souza, P., Imperatriz-Fonseca, V. (2019). Applications of RFID technology on the study of bees. *Insectes Sociaux*, 66, 15–24.

Obeidat, H., Asif, R., Ali, N., Dama, Y., Obeidat, O., Jones, S., Shuaieb, W., Al-Sadoon, M.A., Hameed, K., Alabdullah, A. et al. (2018). An indoor path loss prediction model using wall correction factors for wireless local area network and 5G indoor networks. *Radio Science*, 53(4), 544–564.

Occhiuzzi, C., Parrella, S., Marrocco, G. (2020). Flexible dual chip on-skin RFID sensor for body temperature monitoring. In *2020 IEEE International Conference on Flexible and Printable Sensors and Systems (FLEPS)*, Manchester.

Pan, L., Wang, C., Jin, H., Li, J., Yang, L., Zheng, Y., Wen, Y., Tan, B.H., Loh, X.J., Chen, X. (2020). Lab-on-mask for remote respiratory monitoring. *ACS Materials Letters*, 2(9), 1178–1181.

Parks, A.N., Sample, A.P., Zhao, Y., Smith, J.R. (2013). A wireless sensing platform utilizing ambient RF energy. In *2013 IEEE Topical Conference on Biomedical Wireless Technologies, Networks, and Sensing Systems*, Austin, TX.

Parks, A.N., Liu, A., Gollakota, S., Smith, J.R. (2014). Turbocharging ambient backscatter communication. *ACM SIGCOMM Computer Communication Review*, 44(4), 619–630.

Patwari, N. and Wilson, J. (2010). RF sensor networks for device-free localization: Measurements, models, and algorithms. *Proceedings of the IEEE*, 98(11), 1961–1973.

Pedregosa, F., Varoquaux, G., Gramfort, A., Michel, V., Thirion, B., Grisel, O., Blondel, M., Prettenhofer, P., Weiss, R., Dubourg, V. et al. (2011). Scikit-learn: Machine learning in Python. *The Journal of Machine Learning Research*, 12, 2825–2830.

Pena, D., Feick, R., Hristov, H.D., Grote, W. (2003). Measurement and modeling of propagation losses in brick and concrete walls for the 900-MHz band. *IEEE Transactions on Antennas and Propagation*, 51(1), 31–39.

Peng, Y., Shangguan, L., Hu, Y., Qian, Y., Lin, X., Chen, X., Fang, D., Jamieson, K. (2018). PLoRa: A passive long-range data network from ambient LoRa transmissions. In *Proceedings of the 2018 Conference of the ACM Special Interest Group on Data Communication*, New York.

Pérez-Penichet, C., Hermans, F., Varshney, A., Voigt, T. (2016). Augmenting IoT networks with backscatter-enabled passive sensor tags. In *Proceedings of the 3rd Workshop on Hot Topics in Wireless*. ACM, New York.

Peterson, D.M., Turner, W., Pham, K., Yu, H., Bashirullah, R., Euliano, N., Fitzsimmons, J.R. (2010). A tissue equivalent phantom of the human torso for in vivo biocompatible communications. In *26th Southern Biomedical Engineering Conference SBEC 2010*. Springer, Berlin, Heidelberg.

Pradityo, H.P., Rosyidi, L., Sari, R.F. (2017). Performance evaluation of RSS fingerprinting method to track ZigBee devices location using artificial neural networks. In *2017 International Conference on Information and Communication Technology Convergence (ICTC)*. IEEE, Jeju.

Prata, J.C., Silva, A.L., Walker, T.R., Duarte, A.C., Rocha-Santos, T. (2020). COVID-19 pandemic repercussions on the use and management of plastics. *Environmental Science & Technology*, 54(13), 7760–7765.

Raghavan, A.N., Ananthapadmanaban, H., Sivamurugan, M.S., Ravindran, B. (2010). Accurate mobile robot localization in indoor environments using bluetooth. In *2010 IEEE International Conference on Robotics and Automation*, Anchorage, AK.

RamRakhyani, A.K., Mirabbasi, S., Chiao, M. (2010). Design and optimization of resonance-based efficient wireless power delivery systems for biomedical implants. *IEEE Transactions on Biomedical Circuits and Systems*, 5(1), 48–63.

Ray, P.P. (2017). An IR sensor based smart system to approximate core body temperature. *Journal of Medical Systems*, 41(8), 1–10.

Rayhana, R., Xiao, G., Liu, Z. (2021). RFID sensing technologies for smart agriculture. *IEEE Instrumentation & Measurement Magazine*, 24(3), 50–60.

Saadeh, W., Altaf, M.A.B., Alsuradi, H., Yoo, J. (2017). A pseudo OFDM with miniaturized FSK demodulation body-coupled communication transceiver for binaural hearing aids in 65 nm CMOS. *IEEE Journal of Solid-State Circuits*, 52(3), 757–768.

Sakpere, W.E., Mlitwa, N.B.W., Oshin, M.A. (2017). Towards an efficient indoor navigation system: A near field communication approach. *Journal of Engineering, Design and Technology*, 15(4), 505–527. doi: 10.1108/JEDT-10-2016-0073.

Sample, A. and Smith, J.R. (2009). Experimental results with two wireless power transfer systems. In *2009 IEEE Radio and Wireless Symposium*, San Diego, CA.

Scanlan, J. and Lim, J. (1964). A design theory for optimum broadband reflection amplifiers. *IEEE Transactions on Microwave Theory and Techniques*, 12(5), 504–511.

Schlömer, T., Poppinga, B., Henze, N., Boll, S. (2008). Gesture recognition with a Wii controller. In *Proceedings of the 2nd International Conference on Tangible and Embedded Interaction*. ACM, New York.

Seco, F., Jiménez, A.R., Prieto, C., Roa, J., Koutsou, K. (2009). A survey of mathematical methods for indoor localization. In *2009 IEEE International Symposium on Intelligent Signal Processing*, Budapest.

Semtech Corporation (2013). SX1272/3/6/7/8: LoRa modem design guide. Design Guide, Semtech Corporation, Camarillo, CA.

Semtech Corporation (2015). Application note AN1200.22 LoRa modulation basics. Application Note, Semtech Corporation, Camarillo, CA.

Semtech Corporation (2017). Application note AN1200.14 LoRa modulation crystal oscillator guidance. Application Note, Semtech Corporation, Camarillo, CA.

Semtech Corporation (2020). SX1276/77/78/79–137 MHz to 1020 MHz low power long range transceiver. Data Sheet, Semtech Corporation, Camarillo, CA.

Semtech Corporation (2021). LoRa technology basics [Online]. Available at: https://lora-developers.semtech.com/resources/tech-basics/ [Accessed 25 June 2023].

Shajkofci, A. (2021). Correction of human forehead temperature variations measured by non-contact infrared thermometer. *IEEE Sensors Journal*, 22(7), 16750–16755.

Sheikh, M.U., Xie, B., Ruttik, K., Yiğitler, H., Jäntti, R., Hämäläinen, J. (2021). Ultra-low-power wide range backscatter communication using cellular generated carrier. *Sensors*, 21(8), 2663.

Shotton, J., Fitzgibbon, A., Cook, M., Sharp, T., Finocchio, M., Moore, R., Kipman, A., Blake, A. (2011). Real-time human pose recognition in parts from single depth images. In *CVPR 2011*. IEEE, Colorado Springs, CO.

Siachalou, S., Megalou, S., Tzitzis, A., Tsardoulias, E., Bletsas, A., Sahalos, J., Yioultsis, T., Dimitriou, A.G. (2019). Robotic inventorying and localization of rfid tags, exploiting phase-fingerprinting. In *2019 IEEE International Conference on RFID Technology and Applications (RFID-TA)*, Pisa.

Siegel, P.H. (2021). Microwaves are everywhere: RFID-do you know where your pet is? *IEEE Journal of Microwaves*, 1(3), 679–688.

Sigfox (2019). Sigfox [Online]. Available at: https://www.sigfox.com/ [Accessed 25 June 2023].

Singh, A. and Lubecke, V.M. (2011). Respiratory monitoring and clutter rejection using a CW Doppler radar with passive RF tags. *IEEE Sensors Journal*, 12(3), 558–565.

Škiljo, M., Šolić, P., Blažević, Z., Rodić, L.D., Perković, T. (2021). UHF RFID: Retail store performance. *IEEE Journal of Radio Frequency Identification*, 6, 481–489.

Skolnik, M.I. (2008). *Radar Handbook*. McGraw Hill, New York.

Smith, J.R., Sample, A.P., Powledge, P.S., Roy, S., Mamishev, A. (2006). A wirelessly-powered platform for sensing and computation. In *UbiComp 2006: Ubiquitous Computing: 8th International Conference, UbiComp 2006*, Orange County, CA.

Soontornpipit, P., Furse, C.M., Chung, Y.C. (2004). Design of implantable microstrip antenna for communication with medical implants. *IEEE Transactions on Microwave Theory and Techniques*, 52(8), 1944–1951.

Souri, A., Hussien, A., Hoseyninezhad, M., Norouzi, M. (2022). A systematic review of IoT communication strategies for an efficient smart environment. *Transactions on Emerging Telecommunications Technologies*, 33(3), e3736.

Stein, W., Aleksieieva, A., Roehr, S., Vossiek, M. (2014). Phase modulated 61 GHz backscatter transponder for FMCW radar-based ranging. In *GeMiC 2014, German Microwave Conference*. IEEE, Aachen.

Steinhart, J.S. and Hart, S.R. (1968). Calibration curves for thermistors. *Deep Sea Research and Oceanographic Abstracts*, 15(4), 497–503.

Stockman, H. (1948). Communication by means of reflected power. *Proceedings of the IRE*, 36(10), 1196–1204.

Strobel, A., Carlowitz, C., Wolf, R., Ellinger, F., Vossiek, M. (2013). A millimeter-wave low-power active backscatter tag for FMCW radar systems. *IEEE Transactions on Microwave Theory and Techniques*, 61(5), 1964–1972.

Susnea, I., Dumitriu, L., Talmaciu, M., Pecheanu, E., Munteanu, D. (2019). Unobtrusive monitoring the daily activity routine of elderly people living alone, with low-cost binary sensors. *Sensors*, 19(10), 2264.

Talla, V., Hessar, M., Kellogg, B., Najafi, A., Smith, J.R., Gollakota, S. (2017a). Lora backscatter: Enabling the vision of ubiquitous connectivity. *Proceedings of the ACM on Interactive, Mobile, Wearable and Ubiquitous Technologies*, 1(3), 1–24.

Talla, V., Kellogg, B., Gollakota, S., Smith, J.R. (2017b). Battery-free cellphone. *Proceedings of the ACM on Interactive, Mobile, Wearable and Ubiquitous Technologies*, 1(2), 1–20.

Tao, Q., Li, Y., Zhong, C., Shao, S., Zhang, Z. (2021). A novel interference cancellation scheme for bistatic backscatter communication systems. *IEEE Communications Letters*, 25(6), 2014–2018.

Tegou, T., Kalamaras, I., Votis, K., Tzovaras, D. (2018). A low-cost room-level indoor localization system with easy setup for medical applications. In *2018 11th IFIP Wireless and Mobile Networking Conference (WMNC)*. IEEE, Prague.

TekWiki (2025). GaAs Tunnel Diode for switching applications, metal-ceramic case [Online]. Available at: https://w140.com/tekwiki/wiki/User:Peter/Russian_tunnel_diodes.

Teo, A.J., Mishra, A., Park, I., Kim, Y.-J., Park, W.-T., Yoon, Y.-J. (2016). Polymeric biomaterials for medical implants and devices. *ACS Biomaterials Science & Engineering*, 2(4), 454–472.

Teran, C., Torrez-Llanos, J., Teran-Miranda, T., Balderrama, C., Shah, N., Villarroel, P. (2012). Clinical accuracy of a non-contact infrared skin thermometer in paediatric practice. *Child: Care, Health and Development*, 38(4), 471–476.

Thomas, S.J., Besnoff, J.S., Reynolds, M.S. (2012). Modulated backscatter for ultra-low power uplinks from wearable and implantable devices. In *Proceedings of the 2012 ACM Workshop on Medical Communication Systems*, New York.

Thread Group (n.d.). Thread [Online]. Available at: https://www.threadgroup.org/ [Accessed 25 June 2023].

Tian, Z., Wang, Z., Li, Z., Zhou, M. (2019). RTIL: A real-time indoor localization system by using angle of arrival of commodity WiFi signal. In *2019 11th International Conference on Wireless Communications and Signal Processing (WCSP)*. IEEE, Xi'an.

Tiusanen, M.J. (2013). Soil scouts: Description and performance of single hop wireless underground sensor nodes. *Ad Hoc Networks*, 11(5), 1610–1618.

Tuset-Peiró, P., Anglès-Vazquez, A., López-Vicario, J., Vilajosana-Guillén, X. (2014). On the suitability of the 433 MHz band for M2M low-power wireless communications: Propagation aspects. *Transactions on Emerging Telecommunications Technologies*, 25(12), 1154–1168.

Ublox (n.d.). XPLR-AOA-1 kit. Direction finding explorer kit with NINA-B4 Bluetooth 5.1 module [Online]. Available at: https://www.u-blox.com/en/product/xplr-aoa-1-kit [Accessed 17 June 2023].

Van Huynh, N., Hoang, D.T., Lu, X., Niyato, D., Wang, P., Kim, D.I. (2018). Ambient backscatter communications: A contemporary survey. *IEEE Communications Surveys & Tutorials*, 20(4), 2889–2922.

Varshney, A., Harms, O., Pérez-Penichet, C., Rohner, C., Hermans, F., Voigt, T. (2017). Lorea: A backscatter architecture that achieves a long communication range. In *Proceedings of the 15th ACM Conference on Embedded Network Sensor Systems*, New York.

Varshney, A., Soleiman, A., Voigt, T. (2019). Tunnelscatter: Low power communication for sensor tags using tunnel diodes. In *The 25th Annual International Conference on Mobile Computing and Networking*. ACM, New York.

Vasisht, D., Zhang, G., Abari, O., Lu, H.-M., Flanz, J., Katabi, D. (2018). In-body backscatter communication and localization. In *Proceedings of the 2018 Conference of the ACM Special Interest Group on Data Communication*, New York.

Velluzzi, F., Fois, A., Dell'osa, A., Tocco, F., Bertetto, A., Serra, C., Melis, L., Bertelli, U., Concu, A. (2021). Mechatronic face mask anti covid-19 to remotely record cardiorespiratory variables in farm's workers engaged in jobs at high risk of infection. *International Journal of Mechanics and Control*, 22(1), 61–76.

Vena, A., Perret, E., Tedjini, S. (2011). Chipless RFID tag using hybrid coding technique. *IEEE Transactions on Microwave Theory and Techniques*, 59(12), 3356–3364.

Venguer, A., Medina, J., Chávez, R., Velázquez, A., Zamudio, A., Il'in, G. (2004). Theoretical and experimental analysis of resonant microwave reflection amplifiers. *Microwave Journal*, 47(10), 80–93.

Venturino, L., Grossi, E., Lops, M., Johnston, J., Wang, X. (2022). Radar-enabled ambient backscatter communication. *arXiv:2208.07061*.

Vinitha, A., Rukmini, M., Dhirajsunehra (2022). Secure and energy aware multi-hop routing protocol in WSN using Taylor-based hybrid optimization algorithm. *Journal of King Saud University-Computer and Information Sciences*, 34(5), 1857–1868.

Vougioukas, G. and Bletsas, A. (2017). 24μ Watt 26m range batteryless backscatter sensors with FM remodulation and selection diversity. In *2017 IEEE International Conference on RFID Technology & Application (RFID-TA)*, Warsaw.

Vougioukas, G. and Bletsas, A. (2018). Switching frequency techniques for universal ambient backscatter networking. *IEEE Journal on Selected Areas in Communications*, 37(2), 464–477.

Wang, A., Iyer, V., Talla, V., Smith, J.R., Gollakota, S. (2017). FM backscatter: Enabling connected cities and smart fabrics. In *Proceedings of the 14th USENIX Symposium on Networked Systems Design and Implementation (NSDI '17)*, Berkeley, CA.

Wang, J., Dhanapal, R.K., Ramakrishnan, P., Balasingam, B., Souza, T., Maev, R. (2019). Active RFID based indoor localization. In *2019 22th International Conference on Information Fusion (FUSION)*. IEEE, Ottawa, ON.

Wang, Q., Chen, S., Zhao, J., Gong, W. (2021). Rapidrider: Efficient WiFi backscatter with uncontrolled ambient signals. In *IEEE INFOCOM 2021–IEEE Conference on Computer Communications*, Vancouver.

Wentz, C.T., Bernstein, J.G., Monahan, P., Guerra, A., Rodriguez, A., Boyden, E.S. (2011). A wirelessly powered and controlled device for optical neural control of freely-behaving animals. *Journal of Neural Engineering*, 8(4), 046021.

Wi-Fi Alliance (2020). Wi-Fi 6 (802.11ax) technology [Online]. Available at: https://www.wi-fi.org/discover-wi-fi/wi-fi-6 [Accessed 25 June 2023].

Wolcott, R.W. and Eustice, R.M. (2014). Visual localization within lidar maps for automated urban driving. In *2014 IEEE/RSJ International Conference on Intelligent Robots and Systems*. IEEE, Chicago, IL.

Xiao, J., Zhou, Z., Yi, Y., Ni, L.M. (2016). A survey on wireless indoor localization from the device perspective. *ACM Computing Surveys (CSUR)*, 49(2), 1–31.

Xie, S., Hu, Y., Wang, Y. (2014). Weighted centroid localization for wireless sensor networks. In *2014 IEEE International Conference on Consumer Electronics-China*, Shenzhen.

Xu, Z. and Gong, W. (2021). Homoscatter: Towards efficient connectivity for ZigBee backscatter system. *arXiv:2111.02140*.

Xu, Z. and Gong, W. (2023). BumbleBee: Enabling the vision of pervasive ZigBee backscatter communication. In *2023 IEEE International Conference on Pervasive Computing and Communications (PerCom)*, Atlanta, GA.

Xu, X., Karis, A.J., Buller, M.J., Santee, W.R. (2013). Relationship between core temperature, skin temperature, and heat flux during exercise in heat. *European Journal of Applied Physiology*, 113(9), 2381–2389.

Xu, S., Zhou, H., Wu, C., Huang, C.-M., Moon, S. (2018). Spatial signal attenuation model of active RFID tags. *IEEE Access*, 6, 6947–6960.

Xu, Y., Shmaliy, Y.S., Li, Y., Chen, X., Guo, H. (2019). Indoor INS/LiDAR-based robot localization with improved robustness using cascaded FIR filter. *IEEE Access*, 7, 34189–34197.

Yaici, K. (2020). IoT forecast: Connectivity management platforms 2019–2028. *IoT Platforms and Technology* [Online]. Available at: https://www.analysysmason.com/research/content/regional-forecasts-/iot-cmp-forecast-rdme0/.

Yamakage, M. and Namiki, A. (2003). Deep temperature monitoring using a zero-heat-flow method. *Journal of Anesthesia*, 17(2), 108–115.

Yang, Z., Wu, C., Liu, Y. (2012). Locating in fingerprint space: Wireless indoor localization with little human intervention. In *Proceedings of the 18th Annual International Conference on Mobile Computing and Networking*. ACM, New York.

Yang, C., Gummeson, J., Sample, A. (2017). Riding the airways: Ultra-wideband ambient backscatter via commercial broadcast systems. In *IEEE INFOCOM 2017 – IEEE Conference on Computer Communications*, Atlanta, GA.

Yang, B., Guo, L., Guo, R., Zhao, M., Zhao, T. (2020). A novel trilateration algorithm for RSSI-based indoor localization. *IEEE Sensors Journal*, 20(14), 8164–8172.

Yassin, A., Nasser, Y., Awad, M., Al-Dubai, A., Liu, R., Yuen, C., Raulefs, R., Aboutanios, E. (2016). Recent advances in indoor localization: A survey on theoretical approaches and applications. *IEEE Communications Surveys & Tutorials*, 19(2), 1327–1346.

Yen, C.-C., Gutierrez, A.E., Veeramani, D., Van Der Weide, D. (2007). Radar cross-section analysis of backscattering RFID tags. *IEEE Antennas and Wireless Propagation Letters*, 6, 279–281.

Youssef, M., Youssef, A., Rieger, C., Shankar, U., Agrawala, A. (2006). Pinpoint: An asynchronous time-based location determination system. In *Proceedings of the 4th International Conference on Mobile Systems, Applications and Services*. ACM, New York.

Z-Wave Alliance (2017). Z-wave specification [Online]. Available at: http://z-wavealliance.org/. [Accessed 25 June 2023].

Zafari, F., Gkelias, A., Leung, K.K. (2019). A survey of indoor localization systems and technologies. *IEEE Communications Surveys & Tutorials*, 21(3), 2568–2599.

Zhang, P., Rostami, M., Hu, P., Ganesan, D. (2016a). Enabling practical backscatter communication for on-body sensors. In *Proceedings of the 2016 ACM SIGCOMM Conference*. ACM, New York.

Zhang, P., Bharadia, D., Joshi, K., Katti, S. (2016b). Enabling backscatter communication among commodity WiFi radios. In *Proceedings of the 2016 ACM SIGCOMM Conference*. ACM, New York.

Zhang, P., Bharadia, D., Joshi, K., Katti, S. (2016c). Hitchhike: Practical backscatter using commodity WiFi. In *Proceedings of the 14th ACM Conference on Embedded Network Sensor Systems CD-ROM*, New York.

Zhang, P., Josephson, C., Bharadia, D., Katti, S. (2017). Freerider: Backscatter communication using commodity radios. In *Proceedings of the 13th International Conference on Emerging Networking EXperiments and Technologies*. ACM, New York.

Zhang, M., Chen, S., Nayak, A., Gong, W. (2020a). Enabling multi-channel backscatter communication for Bluetooth low energy. In *ICC 2020–2020 IEEE International Conference on Communications (ICC)*, Dublin.

Zhang, M., Zhao, J., Chen, S., Gong, W. (2020b). Reliable backscatter with commodity BLE. In *IEEE INFOCOM 2020 – IEEE Conference on Computer Communications*, Toronto, ON.

Zhang, M., Chen, S., Zhao, J., Gong, W. (2021). Commodity-level BLE backscatter. In *Proceedings of the 19th Annual International Conference on Mobile Systems, Applications, and Services*. ACM, New York.

Zhao, L., Wang, H., Li, P., Liu, J. (2017). An improved WiFi indoor localization method combining channel state information and received signal strength. In *2017 36th Chinese Control Conference (CCC)*. IEEE, Dalian.

Zhao, J., Gong, W., Liu, J. (2018a). Spatial stream backscatter using commodity WiFi. In *Proceedings of the 16th Annual International Conference on Mobile Systems, Applications, and Services*. ACM, New York.

Zhao, J., Gong, W., Liu, J. (2018b). X-tandem: Towards multi-hop backscatter communication with commodity WiFi. In *Proceedings of the 24th Annual International Conference on Mobile Computing and Networking*. ACM, New York.

Zhou, Y., Law, C.L., Xia, J. (2011). Ultra low-power RFID tag with precision localization using IR-UWB. In *2011 IEEE MTT-S International Microwave Symposium*, Baltimore, MD.

ZigBee Alliance (2017). ZigBee specification [Online]. Available at: https://zigbeealliance.org/ [Accessed 25 June 2023].

Zuo, J., Liu, S., Xia, H., Qiao, Y. (2018). Multi-phase fingerprint map based on interpolation for indoor localization using iBeacons. *IEEE Sensors Journal*, 18(8), 3351–3359.

Index

Other titles from

iSTE

in

Networks and Telecommunications

2025

CHU Tran Tuan, LABIOD Mohamed Aymen, AUGUSTIN Brice, MELLOUK Abdelhamid
Software-defined Infrastructure: A Novel Concept of Advanced Network Infrastructure
(New Generation Networks Set – Volume 4)

DJEACHANDRANE Abhishek, HOCEINE Said, DELMAS Serge, MELLOUK Abdelhamid
Smart Public Safety Video Surveillance System: Innovative Technologies for Homeland Security and Mission-Critical Operations
(New Generation Networks Set – Volume 5)

TIGANE Samir, KAHLOUL Laid, MELLOUK Abdelhamid
Enhancing Stochastic Petri Nets with Reconfigurability: Modeling, Analysis and Performance Evaluation
(New Generation Networks Set – Volume 6)

2024

HORVÁTH András, TELEK Miklós
Phase Type Distributions: Theory and Application
(Stochastic Models in Computer Science and Telecommunication Networks Set – Volume 2)

2023

RAVALIMINOARIMALALASON Toky Basilide, RANDIMBINDRAINIBE Falimanana
Queues Applied to Telecoms: Courses and Exercises

TONG Van Van, SOUIHI Sami, TRAN Hai-Anh, MELLOUK Abdelhamid
Troubleshooting for Network Operators: The Road to a New Paradigm with Encrypted Traffic
(New Generation Networks Set – Volume 3)

2022

BANNOUR Fetia, SOUIHI Sami, MELLOUK Abdelhamid
Software-Defined Networking: Extending SDN Control to Large-Scale Networks
(New Generation Networks Set – Volume 2)

BENAROUS Leila, BITAM Salim, MELLOUK Abdelhamid
Security in Vehicular Networks: Focus on Location and Identity Privacy
(New Generation Networks Set – Volume 1)

2021

LAUNAY Frédéric
NG-RAN and 5G-NR: 5G Radio Access Network and Radio Interference

2020

GONTRAND Christophe
Digital Communication Techniques

PUJOLLE Guy
Software Networks: Virtualization, SDN, 5G and Security (2nd edition revised and updated)
(Advanced Network Set – Volume 1)

2019

LAUNEY Frédéric, PEREZ André
LTE Advanced Pro: Towards the 5G Mobile Network
Harmonic Concept and Applications

TOUNSI Wiem
Cyber-Vigilance and Digital Trust: Cyber Security in the Era of Cloud Computing and IoT

2018

ANDÍA Gianfranco, DURO Yvan, TEDJINI Smail
Non-linearities in Passive RFID Systems: Third Harmonic Concept and Applications

BOUILLARD Anne, BOYER Marc, LE CORRONC Euriell
Deterministic Network Calculus: From Theory to Practical Implementation

LAUNAY Frédéric, PEREZ André
LTE Advanced Pro: Towards the 5G Mobile Network

PEREZ André
Wi-Fi Integration to the 4G Mobile Network

2017

BENSLAMA Malek, BENSLAMA Achour, ARIS Skander
Quantum Communications in New Telecommunications Systems

HILT Benoit, BERBINEAU Marion, VINEL Alexey, PIROVANO Alain
Networking Simulation for Intelligent Transportation Systems: High Mobile Wireless Nodes

LESAS Anne-Marie, MIRANDA Serge
The Art and Science of NFC Programming
(Intellectual Technologies Set – Volume 3)

2016

AL AGHA Khaldoun, PUJOLLE Guy, ALI-YAHIYA Tara
Mobile and Wireless Networks
(Advanced Network Set – Volume 2)

BATTU Daniel
Communication Networks Economy

BENSLAMA Malek, BATATIA Hadj, MESSAI Abderraouf
Transitions from Digital Communications to Quantum Communications: Concepts and Prospects

CHIASSERINI Carla Fabiana, GRIBAUDO Marco, MANINI Daniele
Analytical Modeling of Wireless Communication Systems
(Stochastic Models in Computer Science and Telecommunication Networks Set – Volume 1)

EL FALLAH SEGHROUCHNI Amal, ISHIKAWA Fuyuki, HÉRAULT Laurent, TOKUDA Hideyuki
Enablers for Smart Cities

PEREZ André
VoLTE and ViLTE

2015

BENSLAMA Malek, BATATIA Hadj, BOUCENNA Mohamed Lamine
Ad Hoc Networks Telecommunications and Game Theory

BENSLAMA Malek, KIAMOUCHE Wassila, BATATIA Hadj
Connections Management Strategies in Satellite Cellular Networks

BERTHOU Pascal, BAUDOIN Cédric, GAYRAUD Thierry, GINESTE Matthieu
Satellite and Terrestrial Hybrid Networks

CUADRA-SANCHEZ Antonio, ARACIL Javier
Traffic Anomaly Detection

LE RUYET Didier, PISCHELLA Mylène
Digital Communications 1: Source and Channel Coding

PEREZ André
LTE and LTE Advanced: 4G Network Radio Interface

PISCHELLA Mylène, LE RUYET Didier
Digital Communications 2: Digital Modulations

2014

ANJUM Bushra, PERROS Harry
Bandwidth Allocation for Video under Quality of Service Constraints

BATTU Daniel
New Telecom Networks: Enterprises and Security

BEN MAHMOUD Mohamed Slim, GUERBER Christophe, LARRIEU Nicolas, PIROVANO Alain, RADZIK José
Aeronautical Air−Ground Data Link Communications

BITAM Salim, MELLOUK Abdelhamid
Bio-inspired Routing Protocols for Vehicular Ad-Hoc Networks

CAMPISTA Miguel Elias Mitre, RUBINSTEIN Marcelo Gonçalves
Advanced Routing Protocols for Wireless Networks

CHETTO Maryline
Real-time Systems Scheduling 1: Fundamentals
Real-time Systems Scheduling 2: Focuses

EXPOSITO Ernesto, DIOP Codé
Smart SOA Platforms in Cloud Computing Architectures

LARRIEU Nicolas, VARET Antoine
Rapid Prototyping of Software for Avionics Systems: Model-oriented Approaches for Complex Systems Certification

MELLOUK Abdelhamid, CUADRA-SANCHEZ Antonio
Quality of Experience Engineering for Customer Added Value Services

OTEAFY Sharief M.A., HASSANEIN Hossam S.
Dynamic Wireless Sensor Networks

PEREZ André
Network Security

PERRET Etienne
Radio Frequency Identification and Sensors: From RFID to Chipless RFID

REMY Jean-Gabriel, LETAMENDIA Charlotte
LTE Standards
LTE Services

RIGO Michel
Formal Languages, Automata and Numeration Systems 1: Introduction to Combinatorics on Words
Formal Languages, Automata and Numeration Systems 2: Applications to Recognizability and Decidability

TANWIR Savera, PERROS Harry
VBR Video Traffic Models

VAN METER Rodney
Quantum Networking

XIONG Kaiqi
Resource Optimization and Security for Cloud Services

2013

ASSING Dominique, CALÉ Stéphane
Mobile Access Safety: Beyond BYOD

BEN MAHMOUD Mohamed Slim, LARRIEU Nicolas, PIROVANO Alain
Risk Propagation Assessment for Network Security: Application to Airport Communication Network Design

BERTIN Emmanuel, CRESPI Noël
Architecture and Governance for Communication Services

BEYLOT André-Luc, LABIOD Houda
Vehicular Networks: Models and Algorithms

BRITO Gabriel M., VELLOSO Pedro Braconnot, MORAES Igor M.
Information-Centric Networks: A New Paradigm for the Internet

DEUFF Dominique, COSQUER Mathilde
User-Centered Agile Method

DUARTE Otto Carlos, PUJOLLE Guy
Virtual Networks: Pluralistic Approach for the Next Generation of Internet

FOWLER Scott A., MELLOUK Abdelhamid, YAMADA Naomi
LTE-Advanced DRX Mechanism for Power Saving

FERRANT Jean-Loup *et al.*
Synchronous Ethernet and IEEE 1588 in Telecoms: Next Generation Synchronization Networks

MELLOUK Abdelhamid, HOCEINI Said, TRAN Hai Anh
Quality-of-Experience for Multimedia: Application to Content Delivery Network Architecture

NAIT-SIDI-MOH Ahmed, BAKHOUYA Mohamed, GABER Jaafar, WACK Maxime
Geopositioning and Mobility

PEREZ André
Voice over LTE: EPS and IMS Networks

TERRE Michel, PISCHELLA Mylène, VIVIER Emmanuelle
Wireless Telecommunication Systems

2012

AL AGHA Khaldoun
Network Coding

BOUCHET Olivier
Wireless Optical Communications

DECREUSEFOND Laurent, MOYAL Pascal
Stochastic Modeling and Analysis of Telecoms Networks

DUFOUR Jean-Yves
Intelligent Video Surveillance Systems

EXPOSITO Ernesto
Advanced Transport Protocols: Designing the Next Generation

FOULADGAR Javad
Electrothermics

JUMIRA Oswald, ZEADALLY Sherali
Energy Efficiency in Wireless Networks

KRIEF Francine
Green Networking

PEREZ André
Mobile Networks Architecture

POLER Raúl, CARNEIRO Luis Maia, JASINSKI Thomas, ZOLGHADRI Marc, PEDRAZZOLI Paolo
Intelligent Non-hierarchical Manufacturing Networks

2011

BONALD Thomas, FEUILLET Mathieu
Network Performance Analysis

CARBOU Romain, DIAZ Michel, EXPOSITO Ernesto, ROMAN Rodrigo
Digital Home Networking

CHABANNE Hervé, URIEN Pascal, SUSINI Jean-Ferdinand
RFID and the Internet of Things

GARDUNO BARRERA David, DIAZ Michel
Communicating Systems with UML 2: Modeling and Analysis of Network Protocols

LAHEURTE Jean-Marc
Compact Antennas for Wireless Communications and Terminals: Theory and Design

PALICOT Jacques
Radio Engineering: From Software Radio to Cognitive Radio

PEREZ André
IP, Ethernet and MPLS Networks: Resource and Fault Management

RÉMY Jean-Gabriel, LETAMENDIA Charlotte
Home Area Networks and IPTV

TOUTAIN Laurent, MINABURO Ana
Local Networks and the Internet: From Protocols to Interconnection

2010

BEGAUD Xavier
Ultra Wide Band Antennas

CHAOUCHI Hakima
The Internet of Things

FRIKHA Mounir
Ad Hoc Networks: Routing, QoS and Optimization

KRIEF Francine
Communicating Embedded Systems / Network Applications

2009

CHAOUCHI Hakima, MAKNAVICIUS Maryline
Wireless and Mobile Network Security

VIVIER Emmanuelle
Radio Resources Management in WiMAX

2008

CHADUC Jean-Marc, POGOREL Gérard
The Radio Spectrum

GAÏTI Dominique
Autonomic Networks

LABIOD Houda
Wireless Ad Hoc and Sensor Networks

LECOY Pierre
Fiber-optic Communications

MELLOUK Abdelhamid
End-to-End Quality of Service Engineering in Next Generation Heterogeneous Networks

PAGANI Pascal *et al.*
Ultra-wideband Radio Propagation Channel

2007

BENSLIMANE Abderrahim
Multimedia Multicast on the Internet

PUJOLLE Guy
Management, Control and Evolution of IP Networks

SANCHEZ Javier, THIOUNE Mamadou
UMTS

VIVIER Guillaume
Reconfigurable Mobile Radio System

Printed and bound by CPI Group (UK) Ltd, Croydon, CR0 4YY
13/09/2026
14961012-0001